Two Years and Four Months in a Lunatic Asylum

by Hiram Chase

PREFACE.

I have been urged ever since I left the Asylum, by friends, to write my history of those two unfortunate years, and give it to the public. This I did purpose to do while I was in the asylum, as soon as I left it, while all things would be fresh in my memory. But after leaving that place, and mingling again with the world and with my friends, the very thought of the subject sickened me, for I desired to think and talk as little about the matter as possible.

Besides this, in eighteen months after I left the asylum I entered upon the regular work of the ministry again, and did not wish, while in the effective work of the ministry, to mix with it the history of those two unhappy years, of which I knew, the public had no adequate conception; and which, if I should write out faithfully, would develop facts which many would disbelieve, while others would only laugh at them, as freaks of my insanity, and not as sober truths.

Another reason which has deterred me from giving to the world the history of those two years, is the fact, that a number of inmates of lunatic Asylums in this country have given to the public their views of asylum life, and one especially, who was in the asylum at Utica, and discharged just before I entered it. I could not help noticing the effect these productions produced on society. In many instances the history was read only to laugh, and pity the insanity of the writer. This case referred to, was a lady from Syracuse.

The object she had in view in writing her narrative was evidently lost, excepting the profit she expected to derive from the sale of the work, which I judge could not have been great. She was very unfortunate in writing this narrative; marks of insanity stand out prominently through all the work in the language she uses, in the low scurrilous manner in which she attacks all who differ from her in opinion, her bitterness to the Church, and its ministers, and especially her low ribaldry concerning the doctors of the institution.

On reading this pamphlet, I saw the difficulty attending the writing a narrative of asylum life by a patient, however truthful it might be; for, notwithstanding all the objections that can be raised to the work above referred to, it nevertheless contains many truths of an alarming character--truths which every sane inmate can testify to. And by the way, it must not be supposed that every patient in that institution is insane; far from it; but more of this in its proper place. And though very much, if not all that is related in that pamphlet concerning the institution, is strictly true, yet the manner and spirit in which it is told, detracts very much from its merits.

Considering all these facts, with some others that I need not name, I hesitated, and at one time thought that I would never write one word on the subject. But notwithstanding all the objections that have crowded themselves upon my mind to such an undertaking, I confess I felt myself urged onward to write the facts as they presented themselves to me; and this work I have undertaken, hoping that by this means the public mind may become somewhat, at least, disabused in relation to lunatic asylums in general, and especially in regard to the State Lunatic Asylum at Utica, N.Y.

About three years have elapsed since I left that institution. Since that time I have mingled with society as formerly, have ever since I left preached Christ and the Resurrection as a regular minister in the church of Christ, have done a little worldly business, and am still employed in worldly matters in connection with preaching about every Sabbath. I would state still further that I have been a minister of Christ more than forty years, and more than forty years have been a member of the Troy Conference of the M.E. Church; and I am positive that during that long period no charges or complaints have ever been made against me for immorality, imprudence, or heresy.

I have been thus particular in describing my present standing, to show the public that I have not entered upon this subject as a mad-man, or a man broken down in society. I am not aware that anybody, in or out of the church, looks upon me otherwise than before I went to that institution.

I wish to also state to the public, to the praise of God, that I have not had a sick day since I left the asylum. And what will perhaps appear more strange to the reader, is, that I am prepared to say, and even to prove, that during the two years and four months I was in the institution, I never had a sick day--never lost one meal, but went to the table three times each and every day of the two years and four months; and though over sixty years old when I entered the asylum, I am positive that I never laid down ten minutes upon the bed in the day time during the whole of that time; yet, there were times in which it would have given me great relief from my rheumatic pains, could I have done so, but it was not the good pleasure of the doctors to allow this privilege to me.

I wish to also say, before I enter upon the rather painful work of narrating the events of my captivity, (for I can call it by no other name so appropriate,) that before I went to the asylum as a patient, I was totally ignorant of the character of these institutions. I had never heard them described, except in one instance, and that by a man who was so unfortunate as to be carried there by force by his neighbors, as most patients are carried there. He gave me a most horrible description of his treatment, while in the asylum; how he was dragged by his hair, beaten and bruised, and how he finally made his escape and went home. I heard his sad and tragical tale, but I disposed of it as most men do, by regarding the whole story as imaginary, the effect of a disordered mind, believing that such things could never be tolerated in a Christian country.

It is true that at the time he related this to me, he was sober and in his right mind, and was one of the best of men; yet, I regarded the story at the time, as the wild freak of a disordered mind. I now believe he told me the truth. He afterwards died in the asylum.

To show the general impression on the minds of the outer world on the management of such institutions, and the treatment of their inmates, I was once conversing with a man whose neighbor's wife was an inmate of the Asylum at Utica, and had been for years. I asked him his opinion on the propriety of keeping one so many years in the asylum. He gave it as his opinion that in such public institutions the doctors were the most wise and skillful men in the world; that the nurses and attendants were well skilled in the business; that great care and patience were exercised over the patients, and that no stone was left unturned to soothe and comfort these unfortunate victims of insanity. This certainly was a very charitable view to take of an institution got up professedly for the purpose of relieving that unfortunate class of society. I cheerfully gave my assent to his relation of supposed facts, and went away feeling gratified that we had in our State so noble and so humane an institution.

But the sequel will tell how near right he was in his conclusions, and how near right the public mind is generally, concerning the most of public or State institutions, got up ostensibly for the purpose of relieving suffering humanity. Before my narrative will close, the reader will have the opinion of one, at least, whether State lunatic asylums are a blessing or a curse to our country.

I wish to farther say, before I close my preparatory remarks, that I have no selfish motive to induce me to lay open my experience during those two eventful years. It is not for money, of course, that I do it, for in this respect I shall expect to be the loser; and it certainly cannot be to let the world know that I have been an inmate of a lunatic asylum. I do it for the purpose of opening the eyes of the people of the State of New York, that they may enquire more strictly into the nature and workings of the institutions of benevolence, so called, under their control and patronage; to warn the good people of the State of New York to never send their wives, their children, or any of their dependants to a State institution for the cure of any disease of body or mind, where the patient is confined by bolts and bars by legal sanction, and where the sole power over the patient is vested in one man, whose word is law, and whose commands are as imperious as the Sultan's of Turkey. Such is the fact in relation to the Lunatic Asylum in Utica.

Whatever he orders must be done, and as one of the supervisors once said to me, to show the absoluteness of his word, "That if Dr. Gray should order him to carry me out doors head downwards he should most certainly do it, for his word was law." I replied, "All right; so if the doctor should order you to kill me you would do it." He hesitated a moment and said, "No; I don't think I should do that." The idea advanced, however, was, that the patients must understand that the word of one man is the law of this institution, and whoever comes within its walls must bow to this scepter. And though this institution is under the supervision of eight or nine managers, it is also true, that one man, the superintendent or commander-in-chief, has the sole and undisputed control over all the patients as soon as they are received as patients.

Enough, perhaps, has been written to prepare the way for the particular history of the time I was in that institution, so far at least as my experience and observation is concerned; and though three years have passed away since I left the asylum, yet almost everything that happened within my observation seems to be indelibly written on my mind, so that they are as fresh in my recollection as if they had happened but yesterday.

H. CHASE.

TWO YEARS IN THE ASYLUM.

CHAPTER I

In the spring of 1863, I was appointed by the proper authorities of the church, as the pastor of the M.E. Church in the town of Kinderhook, for the third year, having served that people the two previous years. I commenced my new year in good health and fine spirits; all went on favorably, so far as I knew, until about the first of June, when the first shock which I felt which terminated in my downfall, was but a small affair in itself, and at first affected me very slightly, but continued to wear upon me, until another circumstance, arising from a little gossip in the village of Kinderhook, added to my former trial, threw me into a diseased state of body.

The circumstances were as follows: As I had been in the charge the two previous years, the rules of the church did not admit of my being returned the third year, and yet the official board petitioned to have me returned to them the third year. To effect this, and to make it legal for me to return, some alteration or change must be made in the name of the charge. This was effected in the following manner: This charge contained the villages of Kinderhook and Valatie, lying one mile apart, each having a church, and each having preaching every Sabbath. This charge also embraced the little village of Stuyvesant, near the Kinderhook depot. The first year I had Valatie alone; the second year, Kinderhook, which had been a separate station, was taken in or connected with Valatie, both now making but one charge. To effect my return and make it legal, the authorities at the Conference dropped the name of Valatie off from the minutes, and inserted in its stead Stuyvesant, making the charge now read "Kinderhook and Stuyvesant," instead of

Valatie and Kinderhook.

This change of names was observed by some of the friends in Valatie, and they were highly dissatisfied. I explained the cause, and told them that the name would be restored at the end of the year. This did not satisfy some of them, so the fire was kept up; not that any change was made in the work; each had the same service that they had the previous year. I finally told them I was sorry I had returned to them, as they felt so bad about the change of the name of the charge, as it was done solely that I might return to them. I told them it was not my doings; they had asked for my return, and to effect it this change had to be made.

So matters went on for a month, and I supposed all was quiet, and had never heard a lisp but all were satisfied with me, when all at once one of the official board told me that two or three private members of the church had met to consult on the propriety of having me removed from the charge; mixing a little gossip with this, which was studiously kept from me, until this kind brother revealed it to me. I was not moved by it at first; I knew all the official men of the church were in my favor, and they told me not to mind what these two or three had said. This was the first friction I had ever felt in my ministry.

The leaven continued to work in my mind; my health began to give way. The official board visited me, gave me great encouragement, and offered me money; said I could rest, and they would get the pulpits filled until I was better: they did so. My mind became more and more excited; friends came from a distance to comfort me, but all was in vain; little things were magnified to mountains; I knew that I was unmanned, and could not tell why; I imagined things took place that never existed; my mind took a strange turn; I imagined I was the worst of beings, and that thousands must suffer on my account.

I soon became exceedingly restless; wanted to be constantly on the go; wanted to be constantly doing something, and hardly knew what. I felt in a great hurry to have something done. It is true that I knew at the time what I wanted to do, but when I attempted to do it, I would either find opposition by some one or a strange inability to do what I wanted done. I did not give up preaching until the 28th of June. I shall never forget that day; it was Sabbath; I preached in Kinderhook, and, I think, had the Sacrament; it was a day of great gloominess and trial.

The next day being Monday, my wife took me to Hillsdale to see our friends, hoping a change of place and scenery would help me. But O, how restless I was when I got there; I could not be persuaded to stay any length of time; it seemed as though I must go back; and when I got home I was more wretched than ever. I was sorry I went home. We visited the parish the following week, but none knew where we went; my feelings all seemed a wreck. I did not feel sick during all this time. I laid all my feelings at this time to outward circumstances; I suffered them to prey upon my mind. I had always kept clear of all difficulties; was very tenacious about my standing in society. But I thought I now saw that I was liable to suffer as a minister, and also in my moral character as a Christian; and somehow my hands seemed tied. If I resolved on any particular given course, I seemed to have no power or ability to carry it out. I ceased to write in my diary about the middle of July. If I attempted to write anything I could not find words to say what I wanted to, and if I wrote anything I was not satisfied with it, and would tear it out; so I ceased to write altogether.

About this time I took my room and wished to be alone, and yet I wanted my wife near me all the time, and wanted to talk to her constantly upon the same subject. I knew it was a great annoyance to her, and yet it seemed to me that I could not help it. I knew that I was wearing her out by my course, yet I had no control over myself. It seemed to me that she could help me out of all my troubles, at least I acted so, and yet my judgment told me she could not. I groaned much; my appetite now entirely failed; I did not want to eat for days. Sleep entirely left me, and a night seemed an eternity. I prevailed on my wife to take a separate room to prevent my wearing her out with my groanings. I felt now that I did not want to eat, sleep, or drink anything; my flesh seemed to dry down to my bones. It was at this stage of my condition that I felt that I was the worst being in the world. I shall never forget that I thought Jeff. Davis was a saint compared to me; yet I knew all that passed; my mind was as clear to reason as at this moment, but I viewed everything in a most extravagant light.

It was Sunday, about the first of August, that I lay on my bed; I think some of the family were gone to church; I was in great trouble of mind; all that I ever did that was wrong seemed to rush upon my mind, and though I did not have the consolations of religion to comfort me, as I had been accustomed to, yet I wanted to do all things right, and leave nothing undone that I ought to do. I felt that it was probable that I should not live long, and I wanted to die. At that moment, I thought of some nitrate of silver and corrosive sublimate that I had been using for certain purposes, that I had set away rather carelessly, without labeling. Fearing that some of the family might get hold of them, mistaking them for medicine, I sprung from the bed, took the bottle of nitrate of silver, ran out door with it, made a hole in the ground, and meant to empty the contents of the bottle into this hole; but all at once I thought that some animal might get it and be killed by it. I hesitated, ran back with the bottle, then resolved again to bury it. The day was very hot, and I was running about with this bottle in my hand, undecided what to do with it. At that moment my family came in; wanted to know what I had. I told them; they did not believe me, of course, for I had never told them what was in this bottle they snatched it from my hand, and threw it somewhere, I could not see where. I thought I had as much trouble before as I could bear, but this seemed worse than all the rest put together; I imagined that some animal or human being would get hold of that nitrate of silver and be killed, and I should be charged with their death.

The next day I brought out the corrosive sublimate and meant to have buried it, but my wife snatched it from me and threw it into the cook stove; this, too, alarmed me, fearing some one would be poisoned with it, and even warned them all not to eat the food cooked on the stove, lest they should be poisoned. It will be seen that all these things were evidences that my mind had given way and that I was a prostrated man; yet I knew all that passed.

Boils at this time came out on my face and head; they were very painful; I have no knowledge of ever suffering so much pain before as I did with these boils. At this time the rain fell in torrents, with much thunder and lightning; it rained for many days. This rendered the scene to me much more gloomy and dismal.

My physician now gave me medicine, and after a day or two I felt as well as I ever did in my life; got up, my head feeling clear; dressed and went into the garden, and tried to work a little, but I was too weak to do much; discouragement came over me and I gave it up. Friends had called during the last two weeks, but I had refused to see them; I wanted to be alone. From the middle of August until the 19th, I was feeling much better, and my appetite began to come; medicine had had a good effect.[A]

[A] My bowels had been obstinately constipated for ten or twelve days; when the medicine operated, I was better.

On the 19th of August, my physician with another came in, and I was called up to see them; as I walked out, my physician left the house, leaving the other to converse with me. He commenced conversation; I did not understand his object; my wife told me to ask him about the nitrate of silver and corrosive sublimate, and hear what the doctor would say about it. I told him the story as it was, and asked him if he thought any damage could proceed from it. He said no, he thought not; that it might kill the grass, perhaps, where it was thrown, and that would be all. I thought no more of his call; he left, and I have never seen them since. They went immediately to Hudson, I understood, and got a warrant from the judge to take me to the asylum at Utica.

These doctors were Benson and Talmage; their mission was now ended, and I suppose they calculated they had done a great good to their country. It is not a supposable case that men who can coldly deprive a man of his liberty when he is harmless, would ever enquire after his welfare, or send him a word of comfort; of course I never expected it of these men, and I have no doubt, if the truth could be known, that they would have greatly preferred to have had me die in the asylum than to have had me live and come out again.

The next day, the 20th of August, 1863, about 9 o'clock in the morning, I was called out of my room to dress and take a ride as far as the depot. I rose, dressed and went out. I perceived they seemed in a hurry; I got into the wagon with three men besides myself; these were George Harvey, J. Snyder and Rev. A. Farr. As I got into the wagon and saw my trunk, I enquired where they were going. Mr. Harvey told me I was going to the asylum in Utica.

I have always thought until this day that those three men supposed that what I said and did when I was told where I was going, was a sudden outburst of insanity, but I knew as well what I said, and what I did, as they knew; yet I said some things which I ought not to have said. I knew that I was getting better fast; I knew that I had had a terrible time of it; I had felt much better for a few days past; my mind was not as much agitated as it had been. At a glance I took in the whole scene before me. I saw that I had been deceived; that I was torn from home without my consent; was to be shut up with raving maniacs, and probably to die with them. I saw how cold and unfeeling men could be when a little power was given them; I felt that the world and the church had turned against me. I rose in the wagon in despair and indignation; I said strong things; I knew who had been the chief instruments of my imprisonment. I begged to go anywhere else rather than to Utica; when this was denied me, and I was told by Mr. Snyder to sit down, I announced that I should consider myself no longer a member of the Methodist E. Church; that my connection was dissolved. This was an outburst, it is true, and a foolish one, but I knew what I said, and at the time I meant it. I felt that I was forsaken by God and man; I also confessed that I was a bad man, given over by the Almighty, and had no hope. This was the substance of the confession. This was also wrong; even if it had been true, no one could be benefited by such a confession. I knew what I said and I know too what reply was made by Mr. Farr.

I know that these expressions of mine were marks showing that my mind had been racked. I could not control my mind as usual; yet my memory and reasoning powers were not broken; I ought not to have been sent to an insane asylum.

My attendants soon found that there was no need of fetters or handcuffs to get me to Utica; so one after another fell off, leaving me but with one man, and he not much of a giant. When he told me that he had all the papers in his pocket for my commitment, I made up my mind to be a law and order man, and I have never heard that he had any trouble in getting his patient within the bolts and bars of that humane institution, as some are disposed to call it.[B]

[B] I shall never forget that, while on our way to Utica on the cars, between Schenectady and Utica, Mr. Harvey tried to divert my mind from the subject of going to the asylum. He first referred to the case of Gerrit Smith, who had been in the asylum, to show that it was no disgrace to go there; that did not comfort me. He next called my attention to the case of Major Lee, of Sandy Hill, who had recently died, and to the disposition of his property. I knew he did this to divert my mind; I was indifferent to all this, as I knew what it was done for.

We arrived there the same day, and I was locked up in the third story of the building, with about forty raving maniacs. Others must judge of my feelings when I sat down and looked around me and saw where I was, among entire strangers, and all these disfranchised like myself. One of my first thoughts, after I arrived there, was: "Would to God that I were crazy--so crazy that I could not realize where I am, or what I am, or what will be my future."

But more of this in its appropriate place. I now wish to

appropriate a chapter to a particular subject, viz.: to the manner in which patients are sent to the asylum, and the laws of the State of New York on that subject.

CHAPTER II.

It must not be understood that the same mode of operation is practiced in all cases in sending patients to an insane or lunatic asylum. It must be understood also that we are speaking of a State institution, like the one at Utica.

Some patients are supported in that institution solely by the county to which the patient belongs; others are supported partly by the county and partly by the friends of the patient, or by the patient himself or herself, as the case may be; while others, called private patients, are supported wholly by themselves or by their friends.

When a patient is entered as private, it is not necessary to consult doctors, judges or jurors. Suppose it to be a wife, a husband, or a child. The patient is taken to the asylum, terms of entrance are fixed upon with the superintendent, bonds are given or money in advance as security, and the patient is received. In the most of cases the patient is not consulted in the matter. In some cases, however, the patient is consulted, and consents to go; is made to believe that the asylum is like any other infirmary or hospital, where patients are taken to be nursed and cared for, and cured if possible. In the most of cases, perhaps, there is a kind of dread and horror attending a patient taken to the asylum, and very many go against their will. This opposition is generally attributed to their insanity, and is too generally received as evidence that such an one is a fit subject for a lunatic asylum. Should this dread and fear of going to an asylum be made the criterion by which to determine the sanity or insanity of the patient, I have no doubt but more than three-fourths of the people of the State of New York would be adjudged insane.

Another mode of getting a subject into the asylum--the laws of the State having fixed this mode--is as follows: Two physicians are consulted, and if in their judgment the subject is insane, they so represent the case to the judge of the county, and he issues his order to commit this subject to the asylum, and the order is obeyed. This mode of operation covers a vast number of cases, ranging through all the different grades of what physicians may be pleased to call insanity, from acute mania down through melancholia and epilepsy to the dull, moping, driveling state of idiocy, taking, in its sweep, dotage and childishness of old age.

Here a grand field is opened for operation for designing men and women and for speculation. If the subject begins to be a care and burden to the younger portion of the family--if the subject shows some marks of eccentricity--if the patient discovers marks of dotage in the loss of memory which causes frequent questions on the same subject, and especially if a large property lies between the subject and his or her children or relatives--it is an easy matter, in such cases, and very convenient, to place such subjects in a place away from excitement and care, where they will be well used, and nursed as they could not be at home, and at the same time the family relieved of a great burden. The matter is talked up; the good of the patient is only held out to view; the real cause and reasons for this change are concealed. Doctors are consulted, and by the kind and careful representations of the friends of the patient, the doctors are easily made to believe that the subject is no longer fit to manage his or her affairs, and that ease and quiet would greatly contribute to their comfort--especially if they could be removed away and out of sight of home and their business--and so they come to the conclusion that the lunatic asylum would be the appropriate place for them; and so they all come to the conclusion that it would be best to try it for awhile at least. But to make all safe, an order would be best from the judge; then none could complain that oppression had been practiced.

An order is easily obtained, as the patient perhaps is not to be a county charge, but supported out of his or her own money. I have in my mind at this time a number of such cases, with which I have been acquainted; some of them are now in the asylum; others have died there, as most if not all of this class will do. And why should they not die there? They are not placed there to be cured of old age, nor their state of dotage, nor of a suicidal or revengeful spirit. These were not charged upon them, for they were harmless as children; they were placed there for the relief of other minds, and to lessen the cares of those who owe to them their lives and their tenderest watch-care in their declining years! Will such friends or relatives be anxious to know how the old gentleman or lady fares, away from home, locked up as in a prison, and confined by iron grates and bolts? Do such ones care whether their victim, who has always had his liberty, be locked into a cell at night alone, or whether he is locked in with a half dozen of raving mad men? Will such ones inquire whether he suffers with cold, or whether his food is suited to his appetite, and such as he has been used to at his own full board?

Could the beams of these prison houses speak out, and could the stones cry out of the walls of some of those upper back halls in the asylum at Utica, the revelations of the woes and sufferings of humanity would so shock and astonish the outward world, that instead of classing this institution with the humane and benevolent institutions of the country, it would be classed with those ancient Bastilles which have furnished a history of the most cruel and bloody tragedies ever acted under the sun!

I never conceived or realized, until on my way to Utica under keepers, on what a slender thread hangs the liberty of the people of the State of New York. Only the day before, I was feeling that the spell that had lain upon me for more than a month, was broken; all things began to appear more natural; my appetite became strong, though I was weak in body; I looked haggard, but I believed my system was thoroughly cleansed. I know now, and I knew then, that I understood my own case better than others. I only needed a word of encouragement and comfort to set me all right, instead of censure and cold neglect.

My words of self-reproach, and confession of moral delinquency, had been taken advantage of, to charge me with crimes of which I was never guilty. It is true that I felt that I was a great sinner in general; that I never had done anything as I ought to have done it; yet when asked to define what I meant, and name the particulars in which my great sins consisted, I recollect how dumb and vacant my mind would be, and wondered why I could not frame an answer to their questions.

It finally resolved itself into this, that I had done just nothing at all, all my days, and yet had been supported by the people for doing nothing, and that for this I should be damned.

Yet, I say, that all these thoughts were giving way to a more calm and steady state of mind, instead of that fear and haste which had haunted me for more than a month; I began to be more indifferent also to outward circumstances.

Being in my own house, I thought I could act out my feelings without fear of consequences. I never once thought of the danger of getting into the asylum. I had never thought such a thing possible, for I knew that insanity was never known in one of my family. Could I have had one hint that my restlessness was leading to this, I think I could have prevented it, and should have done so.

But I wish here to enter my protest against the manner that thousands are rushed into the asylum, by those who have no knowledge of asylum life and but little if any knowledge of the philosophy of the human mind. Many have been sent there who had been ill but a few days, and were soon over it, and could they have been left at home a week longer, all would have been right; whereas, by being sent to the asylum, they have been kept there confined for two years--for when once in the asylum, it is no easy matter to get away in a short time, unless they run away.

I know men in the asylum who were thrown in there by their friends, under some peculiar influence, who have been there from six to fifteen years; and they are the same now as when they entered it, not insane, but perhaps a little eccentric, or may entertain some notions on religion or philosophy that are not regarded orthodox. They are in good health, perfectly harmless, and, so far as I could judge, would make better inhabitants than one-fourth of the people that are at large.

The question now arises--"What would you have done to remedy the evil of putting men and women in the asylum that should never go there?" My answer is, that I would so change the laws that two inexperienced quack doctors could not govern the destinies of the people of a whole county. I would first require that those men who are to decide on the fate of their neighbors should be men of experience and discretion, and that there should be at least five of them in a county, chosen by the people for that purpose; I would also require that the patient be brought before a jury of twelve men, who shall decide the matter after the five doctors have examined the patient and given their opinions.

I would require that those five doctors should make themselves acquainted with asylum life; I speak now of State asylums, not private ones. I would have them know how patients are treated, as to medicine, diet, &c. For how can a jury or doctors recommend and decide that the asylum is the proper place for the patient, when they know nothing of its character, only that it is called a State Lunatic Asylum?

I would annihilate that argument so often used to induce the ignorant and the innocent to become willing to go to that den of death! The argument is, that many great and noted men have been inmates of the asylum, such as Gerrit Smith, Esquire W., General B. and Judge C. This was the argument used on me while on my way to the prison. I would go still further. I would require that the managers of such an institution should not leave to one man the destinies of so many hundreds of souls; that they should be required to see for themselves all the internal workings of the institution, that its evils may not become chronic and incurable. I would also require that patients be treated as men and women, and not as dumb beasts, in the manner of doctoring them; that the doctors should conform to the same rules that they would out of the institution in doctoring free agents. That is to say, that, when a patient is cured of a certain disease for which medicine is given, the medicine so given shall be taken off or stopped, and not continued for weeks and months after the end is gained for which the medicine is given. To illustrate what I mean: The doctor orders a certain kind of medicine to a patient; it is a strong tonic, for instance, to give strength and an appetite; the patient takes it three times a day for three months; at the end of three months the patient finds himself well, with a strong appetite, and works hard every day. The patient now says to the doctor, that he feels well, has a good appetite, and thinks the medicine had better be taken off, as it begins to act too heavily upon the system. The doctor replies, "that the medicine must not be taken off; that he must take it as long as he lives, and ask no questions." Would any one, out of such an institution, employ such a physician? Now we all know that anywhere but in a lunatic asylum, medicine is not given except in cases of necessity, and when the object is gained for which the medicine is given, the medicine is taken off or withheld. Is this so in the asylum? Every man in that institution, who is sane enough to know the current events of the asylum, knows this is not so. I am a witness, with hundreds besides me, that medicine once ordered will be continued three times a day for two years, without interruption, and no questions asked the patient by the doctors about the effect of that medicine. I know it was so in my case, and no argument or remonstrance could induce the doctor to even change the medicine. I shall have occasion to say more on this subject in another place. I would have this matter regulated.

I have not a doubt, that if all these matters were fully and rightly investigated and controlled, a much smaller number would be sent to the asylum, and those who were sent would have less reason to complain. I do not mean that that institution or any other could be so conducted that none would complain of ill treatment; this perhaps would be impossible; but it could be so conducted that there would be far less suffering there than now exists.

CHAPTER III.

I will now return to the narrative of my two years in the asylum. I never can forget my feelings when I got out of the carriage and walked up the stone steps and into the centre of that mammoth building. The very thought that I was brought to a lunatic asylum, as a patient, was sufficient to take all the man out of me. I glanced my eyes around upon the massive walls, and high ceilings, and sat down. The doctor came, and my case was introduced to him by my attendant; a very few words passed between the doctor and me; I begged to not be left; I knew not what was before me; I had not formed the least idea of the construction of the building, nor of the manner in which patients were managed.

My attendant, Mr. Harvey, gave me over into the hands of Dr. Gray, the superintendent of the asylum, and seemed to be in haste to get out of my sight; at least it seemed so to me. I asked him how long he was going to leave me here; he replied "perhaps about two months; when your folks get settled they will send for you." But instead of coming to take me away in two months, it was ten months before I heard one word from any person I had ever known before I went to that place, though I often enquired. I finally came to the conclusion that my wife must be dead, or I should have heard from her.

The first intimation I received concerning any of my family, was a visit from my daughter from Illinois, ten months after I entered the institution; this to me was as a visit from an angel from Heaven.

But to return to the thread of my narrative. I said but little to the doctor; my spirits were crushed, and I doubt not but I showed it; I was worn to a skeleton; I was well dressed, excepting one thing, and that was invisible. In my haste in the morning in dressing I left off my suspenders, as I was urged to hurry, and supposed I was only going to take a short ride. I observed this when I got to Utica, and got out of the cars; and having left my porte-monnaie at home with my watch, I had no money to purchase a pair. So I ventured to state the fact to the man that accompanied me to Utica, and asked him to buy me a pair; he looked blank and cold towards me, as though he thought I did not know what I asked for, and made me no reply. I felt grieved; I never doubted but he thought it was a freak of madness that caused me to ask for the suspenders. I thought I would not repeat my request, but often thought that at some subsequent time I would show him that I knew what I asked for, and tell him how I felt when he treated the matter so coldly; but I have never mentioned the subject to him since, and should not have mentioned it now, only to show, that no incident happened at that time, however small, but is still fresh in my recollection.

I bade my friend who took me there good bye, with a heavy heart, and the doctor ordered me taken on to the third hall, which was the third story of the building from the ground floor. There I found about forty patients, the majority of which were very insane. I was afraid as I entered the room; I took a seat on a row of benches fastened to the floor. I have already stated that I at this period wished myself as insane as the rest; I then should not be afraid. I saw that they were reckless, raving, and knocking each other. I looked round among the patients to see if I could see any that looked intelligent and sane; I saw a little old white-headed man that looked the most like a sane man of any on the hall. I approached him and spoke to him; I found his name was Francis; a brother of the editor of the Troy Times; he has since died in the asylum.

A small incident took place a few minutes after I entered the hall, that, though small in itself, was nevertheless most annihilating to my feelings. Mr. Jones, one of the attendants of the hall, approached me and said he must search my pockets. So he went into all my pockets, and as good luck would have it, he found nothing but a few pennies; these he said he must take. I said, "very well, take them." He never mentioned it afterwards to me. I have never doubted but many things are taken from the patients in that way that they never get again. I regard it no less a crime than highway robbery, only more low and cowardly.

It must be understood that the patient's word is not received in evidence if it is contradicted by the attendant. An attendant might take anything from a patient, and if complaint should be made by the patient to the superintendent, the attendant has only to deny it, and then woe to that patient, if the attendant pleases to chastise him.

Patients had better suffer than to reveal anything against an attendant, for suffer he will if he does reveal it to the doctor. I shall, perhaps, have occasion to speak of this hereafter, in relation to the loss of my clothing.

I will here state that as the doctors have a name for every degree and kind of mental derangement, mine was termed by them melancholy--a state of gloominess that some would term hypochondria. I believe none ever charged me with being wild and incoherent in my expressions, and though it is proverbial in the institution, among the patients and attendants, that if a man says he is not insane it is a sure sign that he is, so in consequence of this saying, I was careful to say nothing about my own mental condition, only to ask the attendants and doctors, at times, whether they observed marks of insanity in me.

I once asked one of the doctors this question, and he said he did see marks of insanity in me. I expected this answer, for we were disagreeing about the manner in which they were doctoring me. So he gave me to understand that his word was law, and whatever I thought or whatever I said would make no difference; that I must obey his directions, and would often lay his hand on his mouth, thereby indicating to me that I must not speak unless I was spoken to. I pitied the doctor more than I blamed him, for I saw in him positive and decided marks of tyranny that were in his organization: A dark countenance, low built, short neck, a low forehead, not broad, and eyebrows nearly or quite meeting; a peculiar side glance of his eyes, as though he was looking wondrously wise at times; was subject to a low criticism of words; nothing noble and manly about him.

These remarks may seem to the reader not only too severe but uncalled for. I do not doubt but it does so appear, but I cannot help it; and I confess there was nothing I dreaded so much as to see this doctor come on the hall, and it was always a relief to me when he left.

But to return to my first day's experience in the asylum. In two or three hours, supper was announced by the ringing of a bell; all rushed to the end of the hall, and through a doorway into the dining room, where two long tables were set that would seat forty boarders. I was seated between two very insane men--one an Irishman and the other I think a German. The victuals were all on the plates when we sat down, and the tea, or whatever the drink might be called, was already prepared in large pitchers, and poured out in small punch bowls, which were used as a substitute for tea cups and saucers. This was a kind of tea, very weak, prepared with milk and sugar before being turned into the bowls. I tasted it, but it being so different from what I had been used to drinking, as I had never used sugar in my tea, it produced a most sickening influence.

The supper consisted of a couple of pieces of bread, one of wheat, and the other what they called brown or Graham bread--the best I thought I ever had seen--a small piece of butter and a small square piece of gingerbread. As I sat nibbling a little, for I did not eat much the first meal, my Irish companion on the right reached to my plate and took my bread. I looked at him, but he did not notice me; next he reached and took my butter, not seeming to act as though he had done anything out of order. An Irishman on the opposite side of the table reached and took the remainder of my bread and cake, so that when supper was ended, it appeared that I had eaten very heartily, for when I sat down there was enough on my plate to satisfy any hungry man. It was astonishing to see with what rapidity some of those lunatics would devour their food.

When supper was over, one of the attendants came round to every man's plate, took up the knives and counted them, to see if any were missing. This was done to prevent any evil from those who might be suicidal or otherwise evil disposed. At a given signal, all arose and went out. I observed, however, that a number of patients staid in the dining room to help in clearing off the table, washing up, and setting the table again for breakfast.

As soon as this was ended, I heard a sound ringing through the whole length of the hall, "Bed time gents." I thought it very strange, as the sun was yet an hour high. The attendant came to me and told me I must retire. I said, "this is earlier than I am used to retiring."

He made no reply, but led me into a large dormitory, at the end of the hall, containing five beds. One of these was assigned to me; the others were occupied by two Irishmen, and two Americans--one from Saratoga Springs, by the name of Burnham, the other from Hartford, N.Y., whose name I do not now recollect. Burnham and one of the Irishmen were very crazy. The Irishman would get out of bed, wrap himself in his sheet, walk the room, or stand and look out the window, keeping up, in the meantime, an incoherent jingle of words, mixing it with cursings on all Protestants, threatening to scald them to death with hot water; while Burnham would damn him and pour upon him the most bitter curses.

I tried to appease them by flattery. So the night wore away, and in consequence of the novelty of the scene, being locked up in a room with four crazy men, our clothing left out in the hall, the quarreling of my room-mates, with now and then a wild yell from some other apartment, were not very favorable accompaniments to sleep or rest. Sleep entirely departed; I did not feel the least sensation of sleep during the whole night.

No one, unless placed in the same condition with myself, can imagine with what pain and anguish I passed that first night of my captivity. I had already seen that patients were treated more like prisoners than like innocent men and invalids. I had been in bed about an hour, it being now about sundown, when one of the attendants, a gladiatorial looking German, entered the dormitory bearing in his hand a tray of medicine, arranged in rows in little white earthen mugs, each holding perhaps a half of a gill; he came to my bed side and held out one of these mugs to me, and said in broken English, "trink tis." I had seen enough already to know that resistance or remonstrance was of no use, so drank the nostrum but a more nauseous dose I never took. In half an hour more another dose was presented of another kind, I knew not what. I began to think by this time that if a man could live through all this, he must be made of stern stuff.

The morning came, and I was glad to see the light and to get out of that inner prison, where I could have a little wider liberty in walking the hall, which was about two hundred feet long by ten or twelve feet wide, with bed-rooms ranged on each side. After adjusting matters in my sleeping room, making beds and sweeping--as every patient is expected to make his or her own bed, unless unable to do so by physical or mental inability--I was introduced to a wash room. In this room there were barely accommodations to wash by forty patients washing out of about two tin wash dishes, one after another, till all had finished, and then all wipe on about two towels hanging on the wall. No looking glass, combs or brushes were furnished for patients on this hall. I did not see my face in a glass until I left that hall and went to another, which was six weeks from the time I entered the asylum.

Breakfast was announced by the same ringing of the bell. The men were soon in their places; I lingered a little, and was urged forward. I was no sooner in my old place by the side of my agreeable companions, than the Irishman on my right snatched at my bread on my plate; I turned his arm aside, but he seemed determined to seize the whole contents of my plate, which consisted of bread, potato, a piece of cold baked beef and a small piece of butter. Instead of tea, it was called coffee, prepared in the same manner of the tea. I could not drink it at first, but finally by degrees worked myself up to the point. A most wonderful drink is this for sick folks!

As to the diet, so far as I was concerned, all was well enough;

they make use of a vast amount of meat; and it was amusing to hear Dr. Gray philosophise on the utility of the patients eating so much meat. This, perhaps, was all well enough, but in no place but in a lunatic asylum would such doctrines be urged, expecting the people would indorse them, as a diet for invalids, and especially for invalids whose disease is supposed to be mostly of the mind. It is generally supposed, by reasoning beings, that less meat and more variety in lighter food, would be more appropriate to such constitutions as the asylum is made up of. But concerning this matter, I must give no decided opinion; I consider it of minor importance, compared with other things.

The second day had now come; it was Friday, the 21st day of August. I took the medicine in the morning, and after breakfast set myself to learning all I could of the institution by observation. I noticed that in some of the rooms were cribs in which were confined patients by a lid or cover, locked down; these I regarded as men who were not safe to have their liberty and to lie on ordinary beds, and I found this was so. I noticed also that food was carried to these, of a very light nature, as farina or a little soup, and sometimes a small piece of toast. I wondered how men could live on so very little as they seemed to give them, but perhaps they had all that was necessary.

I noticed one among these, of a manly and noble bearing, when he would rise out of his crib; and on inquiry, I found his name to be Maulby, Doctor Maulby, who had been in the institution for many years; and before I left the institution he died there. He was a man, I was informed, of superior talents, and at times was very insane.

In one end of this hall, I observed a large wardrobe or closet, in which all the clothing of the patients was kept for this hall. No patient on this hall is allowed to keep his clothing in his own room; and indeed this is the case with every other hall in the building, except the first hall, which is used mostly for cured patients and the convalescent.

On the gentlemen's side of the house, there are about twelve halls occupied by patients, making in all about three hundred; and as many on the opposite side of the building occupied by females, averaging in all, perhaps, as a general thing, about six hundred.

About nine o'clock in the morning of this my second day in the asylum, I observed a rush of all the patients around a large basket which had been brought out, containing their hats and caps. It was the hour of going out to walk and take the air in the yard, an enclosure attached to the building, of two or three acres, guarded on two sides by the building and on the other two by a high board fence. This yard was beautifully laid out in walks, and covered with grass, trees, and shrubbery.

I supposed I must go out with the rest of the lunatics, so I walked up to take my hat, but I was told I could not go. I could not see the point at that time, but afterwards learned that no patients, when they first come, are allowed to go out until they have been there a number of days. I was glad of this, for I preferred staying in alone to going out with that motley group of maniacs. Not only from this hall did patients go into this yard, but from all the halls, except the first, second and fourth, and sometimes they went from these; and when all these came together, it furnished a most interesting yet ludicrous picture--all the nations of the earth here represented, making a perfect bedlam.

I spent the forenoon as best I could, walking up and down the hall, and sitting alone in my glory; all seemed a blank. In the course of the forenoon Dr. C., who had charge of the north wards of the building, which contained the men's side, came on the hall. He introduced himself to me as the physician of this ward, and took some pains to impress me with the idea that "he was the boss of this shanty," and that his orders must be carried out to the letter. This doctor had charge of the men under Dr. Gray, while Dr. Kellogg had charge of the female department.

This first interview with the doctor made an unfavorable impression upon my mind. I next came to a point in my experience in the institution which added greatly to my fears, and filled me with anguish, and robbed me of all confidence in the attendants, that they had any regard whatever for the feelings and comfort of the patients.

In the afternoon of Friday, my second day in the asylum, I was told by the little Dutchman, the second attendant, to go with him; I followed; he went into the bath-room, carrying a change of my clean underclothes, which they had taken from my trunk; when in the bath-room he locked the door; there stood the bath, about two-thirds full of water, or rather mud and slime, in which ten or twelve filthy maniacs had been scrubbed and washed with soft soap, until the water had become quite thick and disgusting to look upon. He said to me in his broken English, "untress you and kit in dare." I looked at him and said, "am I to bathe in that mud and slush?" he said, "yes, kit in dare quick." I saw I was sold; I was weak in body, the door locked, and though when in my full strength could have thrown him into the bath and held him there; yet now I doubted my ability to vie with him, and besides, I knew he had the power to call to his aid whoever he chose. I did not deliberate long; I threw off my clothes and jumped in, but jumped out as soon as I went in, and called for a towel to wipe off the filth; he refused to give me one, but ordered me to take my cast-off shirt and wipe myself with it. I did so as well as I could, and begged for a clean pail of water to wash myself with, but this was refused.

I made complaint to the first attendant on the hall, but got no satisfaction. I saw the matter was all understood between them; it was done to save time and a little work. There was water plenty, so that each and every man could have had a clean bath; if not, it were far better to not bathe at all, than to bathe in a mud hole. But the laws must be obeyed to make each and every patient bathe once a week. I knew if I complained to the doctor, it would be no better, for he would either justify the course, or the attendants would deny that such an event ever took place, and I alone would be the sufferer.

I did, however, before I left the institution, lay this matter, with some other things, before Dr. S., a fine humane man who was in the institution for a year before I left. He believed my story and reprobated the course. I only wished at the time that those who forced me into such measures had been obliged to bathe in the same slough hole.

Such attendants are men that never went in good society. I can say as Job said of those who taunted him in his affliction, that they were men that he, before he was cast down, would not have associated with his dogs; yet, now they ridiculed him when he was in trouble. So say I; these are men that now, and before I went to the asylum, I should have been ashamed to associate with, but having a little power, they humbled me, and in fear I obeyed them, yet I despised them, and I cannot forget them.

CHAPTER IV.

The first Sabbath came the 23d of August. I had seen nothing of the institution as yet, only what I had seen from this hall. I could only look out of a north window, and see the hills afar off, the valley of the Mohawk stretching east and west as far as the eye could reach; could see the cars passing up and down the valley, and the canal, with its loaded crafts slowly but constantly passing by. I could also see fine carriages constantly passing by, going in and out of the city. I could also see the beautiful lawn lying at my feet, and stretching away to the street passing out of the city. While I stood at my window and saw all this, and then turned and looked at myself, shut up and confined with bars and bolts, I then began to think that I could now conceive how those poor creatures felt whom I had often seen crowding to prison windows to catch a glimpse of passers-by, through their iron grates.

I recollect, while thus employed and thus philosophising, of crying out, that "my life is a failure." I had never realized before the sweets of liberty, and finally came almost to the conclusion that I must have committed some crime, or I never should have been thus confined and shut out from society; yet I had no knowledge that I had violated the law in any sense.

Yes, this was a lonely Sabbath; yet I felt that while I remained in that institution, I had no desire to go out or to form any acquaintances. I could not get rid of the idea that the whole process of proceedings in putting me into the asylum was deception from end to end. First, they were deceived as to the cause of my trouble; secondly, they were deceived in regard to my real condition. I did not wish to look any man in the face, outside of the asylum, for the reason that I supposed all within its walls were regarded as insane and unfit to mingle in society.

I learned that there was service in the chapel that evening, but nothing was said to me about attending; and I did not mention it, for fear I should be denied the privilege of attending.

A day or two more passed away, and I had not, as yet, put off my best clothes. I was thinking of it, and then I thought again--"Why should I care about the future? And if I lay off this suit I shall never see it again." These were thoughts that came into my mind; and I thought I might as well wear out my best clothes as to let others have them.

While these thoughts were revolving in my mind, Mr. Jones, the attendant, came to me and said--"You had better lay off that suit of clothes, and put on a poorer one, to wallow on the hall in." So I made the change, as I had a number of poorer suits in my trunk. This suit that I laid off was a very fine one and valuable. Time went on, and in about six weeks I was removed to the fourth floor. This was a short hall on the first floor, extending west from the main building; but the same suit of clothes that I laid off, a few days after I entered the asylum, I never saw again. I was never fully satisfied what became of them.

The State Fair was held in Utica that fall, and I was invited to ride on to the grounds, with others, in an omnibus. I did not care to go, yet I did not think it best to refuse; I consented and called for my coat; a coat was brought me, but it was not mine; it was much smaller, shorter sleeves, and much worn; it was not worth ten dollars; mine was worth thirty.

I made this known, but all the satisfaction I got was to be told that

I was mistaken. I soon called for my pants and vest which belonged to that coat, and was told by the attendant on that hall, that I never had such a pair of pants and vest as I described--a fine pair of doeskins, and a satin vest; and he told me if I persisted in it, he would report me to higher authority; he even threatened me. I knew I was right, yet I became afraid of my safety, as this attendant on the fourth hall was an old Irishman who had been a sailor, whose principles were very bad; he was not a man of truth or honesty; so I was obliged to let the matter drop. I once thought of stating the matter to Dr. Gray, but the attendants put on their veto, and I let it rest, but have never doubted but this same old Irishman had my vest, for I am sure I saw him wear it. As to the pants, I never saw them again. I know I am not mistaken about the coat, vest and pants; I got an old coat in its stead which I still keep to show.

Cold weather soon came on, and I was thinly clad. I missed my thick pants, and though I had a good shawl, which I kept my eye upon, yet I had no overcoat. I one day said to the supervisor that I wished I had my overcoats from home; that I had two at home--one new and a very fine one, the other a coarse one, but a good coat for common wear.

A very few days after this my coats both came; I knew them well, by special marks. The best one was taken and put away; the other I was allowed to keep in my room to throw on when we went out in the field. It was not long before I called for my best overcoat, as I was going to walk out. A coat was brought me, but on examination, I found it was not my coat; it was much smaller, cut in a different fashion; was not the same kind of cloth; yet it was a black coat, and had a velvet collar like mine; mine was worth at that time fifty dollars; this was not worth twenty. I have never worn the coat much since. I got me a new one and keep this also to exhibit, to show that I am not mistaken about the clothing.

My hat was also changed for one much poorer; this might have been done by accident. A new black silk cravat was taken, and an old one given me in its stead. Now all these things might have been done through mistake, and not by design, yet, I have never doubted but all was done by design; knowing the attendants, I am obliged to come to this conclusion.

It will be observed that, for the sake of giving a history of my lost clothing, the reader was brought down from the third to the fourth floor; as I had not proceeded through an entire week with my history of that hall, we will now return to that narrative. I had been there about a week when I was permitted to go out in the yard with the patients; and in walking in the yard, I soon became acquainted with men from other halls, with whom I could converse, and I found, on comparison, that those on the third were not as sane as many from other halls; indeed, there were none on that hall that could converse rationally for any length of time; yet I did not desire to change my place by being removed to another floor. After being there about a month, however, the doctor hinted to me that I was to be removed to some other floor. This I somehow dreaded, not knowing where I was to be sent, and not knowing the difference between one hall and another; I begged to stay where I was, choosing the sufferings I then had, to those I knew not of.

After being there about two weeks, I one day said to the attendant, that I wished him to understand that if my plate was always found emptied of its contents, at the close of every meal, it was not because I had eaten it all. I then told him it was very annoying to me to have men snatching my food from my plate every chance they could get, and that I was obliged to guard my plate in order to get enough to eat, and the moment I finished, my plate was immediately swept clean of all it contained. He said I should sit there no longer; so he removed me to the table where he sat, and placed me by his side, and I sat there until removed to another hall.

As I have said, I was on this third hall about six weeks. I have noticed but few incidents connected with this hall, not because I could not, but because I wish to make my narrative as short as possible. Should I record all the thrilling and ludicrous incidents which happened upon this hall, and others during my stay there, they would fill an octavo of a thousand pages. My object is not to give a history of the institution, but simply my own narrative, noticing, perhaps, now and then, a circumstance which may fall in my way concerning other patients; and while I am on this subject, I will simply say, that I made the acquaintance of a number of gentlemen in that institution whose names I remember with pleasure, and should perhaps make mention of them if I thought it would be pleasing to them, but knowing the delicacy of such a subject, I shall forbear making mention of any except those who I know cannot be affected by it.

I was now placed upon the fourth hall, and assigned to a room containing three beds; this was about the first of October. The inmates of this room were more agreeable than on the third floor, though one of them, at times, was very annoying. He would be up and down all night; would disarrange all the clothes of his bed; would scold and worry, and complain of ill treatment, if any one attempted to assist him; until at length he was removed on to some other hall and died there.

From this hall I was suffered to walk out with other patients, guarded by attendants. We would sometimes walk a mile through the back fields attached to the institution. I shall never forget that the first day I entered this hall, I saw, walking the hall, a delicate, well dressed, fine looking gentleman, of middle age and very long beard. There seemed to be an air of aristocracy about him that attracted my attention, and led me to inquire who he was. I found he was from Albany; that his name was Root; they called him Colonel Root. He had done business in Albany; married there into a good family and rich. He lived rather too fast to suit his friends, in traveling through Europe and America, and drinking wine and brandy, so they threw him into the asylum. No one could detect in him any marks of insanity; but the way he would curse his friends for running him into that institution, was a caution. He was not the most gentle and docile patient to manage in the whole institution. Being a private patient, he had what is called his extras in food. He was often changed from one hall to another, until, running down rapidly in health, he died on the sixth hall, long before I left the institution.

I liked the fare better on this hall than on the third; it was a short hall, containing about twenty patients. I soon discovered that on this hall were a good many invalids; I have seen as many as ten confined to their beds on this hall at once; I regarded it a kind of hospital. There was a hospital attached to the institution, but I found, of late, it had not been used much for that purpose; that the sick were allowed to remain on the halls with the well. This I regarded an improvement.

At this time, frequent changes were made on the halls in attendants; it was war time, and young men were called into the field; I suppose they had to take such as they could get. A young man came on to the fourth hall, as first attendant, soon after I entered it, by the name of John Subert; a young man of a good deal of self-conceit; was very ignorant withal, and evidently felt that he was highly promoted in having a kind of charge over a few poor inmates of a lunatic asylum.

Doctor Gray is the sole superintendent of the asylum. He has generally three physicians under him, who watch over the wants of the patients, and prescribe for them. Next comes a supervisor, who takes the general charge of four or five halls, and is at the same time an attendant on one of these four or five. This John Subert was an attendant on the fourth hall, under a supervisor; he was, in fact, nothing more nor less than a servant waiter; yet he sometimes assumed a good deal of authority. He at one time called me to come and sit down by his side, and began to talk to me very gravely, and told me whenever I got into any trouble and wanted anything, to come to him and he would give me good advice. This, certainly, would have been very kind, had it come from Doctor Gray or even from a supervisor; but coming from a waiter, and a young man not much over twenty, and one so ignorant that he could not converse intelligently five minutes on any subject, and withal very wicked, using much profane language, the idea of his giving me good advice was most ludicrous.

I once asked this young gentleman for a coverlid, as the weather was getting cold. He brought me an old straw bed tick, very dirty. I looked at it and then at him, and asked him what he meant, to offer me that dirty bed tick for a covering. I saw he was mad. He said I was the damnedest man he ever saw; would sew me up in the tick. He then asked me if he should knock me down. I told him yes, if he pleased. He said he thought he would not begin with me, as he had never knocked a man down. I have never doubted but it was best that he did not knock me down, or attempt it, for I had regained my strength at that time.

And here I am happy to say that during the two years and four months that I was in the institution, I never received a blow from attendant or patient, while many were knocked headlong by both patients and attendants. I was always on the watch to keep out of the way of danger, and when I found an ill-natured patient, or an ill-natured attendant, had as little to say to them as possible.

It is true, that there are times that a man will pass through scenes that will stir his blood, that perhaps he would not let pass unnoticed out of that place; yet, I found the best way to get along, was to bear all things with a kind of stoicism.

I can never forget a small circumstance which happened on this hall. After I had eaten all that I desired, John Subert presented me with a bowl of soup which he had left. I hesitated; told him that I did not need it. He said I should eat it; to save trouble I ate what I could, and stopped; he ordered me to eat the rest, and said I should eat it. I was in a strait; I felt that I could not swallow another spoonful; he threatened; I ate a spoonful or two and stopped; found it impossible to swallow any more. At this point I felt unmanned; I groaned bitterly; I felt that I had rather die than be governed by such a gladiator. I knew he did it only to show his authority. I never knew why he took such a course with me. Had I refused to eat my regular meals, as some did refuse, and had shown a suicidal spirit to starve myself, as some did, then the case would have been altered, and the attendant would have been justified in forcing me to eat. But I was well and hearty; my appetite was craving, caused by the medicine forced down me daily, and I found that I generally ate more than was for my good; yet I did not eat more than other patients; it was thought I did not generally eat as much.

At another time, they had molasses and some kind of pudding as a dessert. I ate all I wanted and moved back; he had ate and left a quantity of molasses and pudding; he moved it before me and ordered me to eat it; molasses I never eat unless obliged to; I tried to beg off, but he was inflexible; I considered the matter and complied; I thought it better to eat his leavings than to have war at the table. I considered that he was a low-bred wretch, and a man of no principle. I have often wondered if he would not like to see me now, and talk up these matters, and show me that it would be best for me to ask his advice, and to eat his leavings. I have no doubt but he would deny that these things ever happened. I would deny them if I were him. This is the way such men get out from such charges. They have been in the habit of abusing patients, and when charged with the wrong, deny it to the doctor, charging it to the insanity of the patient. Many other small matters in themselves might be related that will be passed over, which would be very trying to a man of good breeding.

When the patients of that institution can be used as patients should be, and not as criminals, prisoners or slaves, then, and not till then, will it become a blessing to the State of New York instead of a curse.

I remained on the fourth hall until about the first of December, when I was removed to the first hall. I begged with all my skill to stay on the fourth hall through the winter, but all was in vain. The reasons why I wanted to stay on the fourth hall were, that it was warmer, and I did not wish to become a gazing stock for the multitude of visitors who daily flocked to the asylum, take a walk through the first hall, gaze on the patients as they would look upon wild animals in a managerie, and then depart. I found the arrangement on the fourth hall for bathing as it should be; each man had his bath by himself of clean water. This became a luxury rather than a dread, as upon the third floor. It is, however, due to Mr. Jones, the attendant on the third, to say that after two or three of the first baths I took there, he gave me clean water, and always used me like a gentleman. The little Dutchman who gave me my first bath, seemed to shun me after I had learned the ropes a little better.

My medicine was kept up while on the third and fourth halls without interruption three times a day, always just before eating; and soon after I came to the fourth hall, another dose was added. This was some kind of spirits; whether it was brandy or some other kind of liquor I do not know; one thing I do know, that it would fly into my head, my face would feel hot and would be as red as fire; it alarmed me at first, and I begged to have it taken off, but it was of no use; perhaps I was foolish in thinking that they meant to make me drunk.

After a week or two this beverage was taken off, and strong beer or porter was substituted; this I hated; I always hated it. I hate it still, though I was made to drink it daily for more than a year, and had I been like some men, I should now be a drunkard; but I have not tasted a drop of ardent spirits or beer since I left the asylum, and never shall, unless it is forced down my throat as it was there. My opinion is, however, that the beer I drank there never injured me, but the other medicine I thought did.

Four months had now gone by since I entered the asylum. I was now on the first floor. This is a spacious hall, two hundred and fifteen feet long, with bed-rooms ranged on each side of it to contain about forty patients. The patients on this hall are mostly those who have been on other halls, and are either cured or convalescent; but few on this hall are ever seen to show marks of insanity.

To judge of the inmates of the asylum, and the workings of the institution by inspecting this hall, would be a deception. All things here are in order, with a fine library and reading room, with bureaus and looking glasses in all the bed-rooms.

When I came on to the first hall, I little understood what was before me; I did not know that I was to remain on this floor for two years longer, confined by iron grates and locks; but such was the fact, though I was in as good health the day I entered it as when I left it, but was not in as good spirits.

For the first three months I occupied a bed in one of the dormitories where there were four beds, and during this time I took care of my own bed, and helped others in the room who were weaker than myself. I had a warm place to sleep, and had the privilege of managing my own clothing. Our cast-off clothing at night were not left out in the hall, as on other halls; yet the patients here are all locked into their rooms at night as on other halls; and instead of retiring at seven o'clock, the time of retiring is half past eight. This to me was a great relief.

This was a very hard winter; the cold was intense; the hall was much colder than any house I had ever been accustomed to during my whole life. My clothing was thinner than I had been accustomed to for thirty years, and we were not allowed to put on an overcoat, or wear a shawl in the house, yet my health was good during the whole winter.

The halls were heated with hot air thrown in through pipes from the engine-house on the opposite side of the court yard. The reading room was always comfortable, but I did not stay in it perhaps six hours during the whole winter.

One circumstance connected with my captivity, I cannot pass over. I found when I arrived at Utica that I had no glasses, and although they were in my trunk, I did not know it, as they had taken charge of my trunk, with all its contents, which I never saw again until it was brought down at the time I finally left. I asked for glasses, that I might occupy my time in reading. This was denied me, and the doctor forbade my reading anything whatever. I thought this a hard case. I could not see the point, inasmuch as I saw others reading who were not half as strong as I was--patients who were confined to their beds had their books and papers to read, while I was waiting on them. I came to the conclusion that it was done to punish me, or to let me know that I must obey orders. So I spent the winter the best I could, straining my eyes to read whenever I could get out of the sight of the attendants, that they might not report me to the doctor; and it was quite remarkable that I could read so well without glasses.

Six months perhaps passed away before I was furnished with glasses. I then took to reading, asking no questions, and no one forbade me. Many a volume, could they speak, in that library, could testify that I searched their contents.

Soon after I went to the first hall I commenced walking out with the patients, accompanied by an attendant. It was our custom to go into the street that leads from Utica to Whitesborough, and follow up that road until we came to the bridge which crosses the canal, a distance of about a mile and a half; here we would stop a few minutes and walk back. This we repeated almost every day through the winter.

After I went on to the first hall, I was a constant attendant at church, either in the chapel in the asylum, or in the city, or both. I generally attended the State Street Methodist Episcopal Church in the city. The services in the chapel are generally Episcopal, Dr. Gibson, of Utica, being the chaplain. I confess that I did not enjoy public worship while in the asylum as I have since I left it.

There was one idea that constantly haunted my mind during the most of the time that I was in the asylum: that was, that I should never get away from that place alive, and this I often expressed to others. This, perhaps, may be regarded by others as a freak of insanity, but I could not help it--I had my own reasons for thinking so.

I never saw the day, from the time I entered the asylum until I left it, but I would have been willing to have crawled upon my hands and feet a hundred miles, and lived on bread and water, could I by that means have got away; and yet I was resolved that I would never run away if I died in the institution. Here I think I was in an error; I have no doubt but a man is justifiable in running away when he sees and knows he is receiving no benefit from staying there. I think it would be very difficult to keep me there again as long under the same circumstances. Why should a man feel any conscientious scruples about leaving a place into which he is forced against his will, especially if he was not sent there for crime?

My conscientious scruples about running away from that place, is to me one of the strongest evidences that I can think of, that my mind, some of the time, was not right.

At the time I was changed to the first hall, I was placed at the table by the side of Dr. Noise, who had been in the institution for three or four years. He carried the keys of the house, went in and out at his will, and served as usher to the asylum. I supposed he was employed by the doctor as an helper in the institution, and had no idea that he was a patient. I observed he acted very independently, and was quite dictatorial.

I did not take pains to make his acquaintance, so I said nothing to him for perhaps two weeks. In the meantime I had learned his history--that he was a patient, and that in consequence of his being an active kind of a man, and being a physician at the same time, and not much, if any, insane, was granted privileges that but few patients enjoyed.

I observed that he wanted a great deal of room at the table, and took it without consulting the convenience of his next neighbor. I found myself much cramped for room, and his course became quite annoying to me. He would spread himself out, lay his arms on the table, slop over his tea and coffee on the table-cloth, throw his meat and potatoes off of his plate if he did not want to eat them, and had very much to say to other patients while eating. On one occasion he took his seat at the table before I did, spread himself out as usual, and laid his arm on my knife. I took my seat as usual, and sat awhile to see if he meant to remove his arm off of my knife. I saw that he did not mean to do so. I did not understand his object, but I soon found it was to draw me out in conversation, as I had not as yet spoken to him, and he began to feel annoyed about it. I at length asked him to remove his arm that I might take my knife.

He turned and looked daggers at me. "What," he said, "have you spoken? I have sat by you two weeks, and you have not spoken to me; you need not try to play possum with me?" "What do you mean," I said, "by playing possum?" He gave his definition of the saying. I then said, "Doctor, I feel under no obligations to you; I know no reason why I should make conversation with you more than others." This offended him; he lifted up his voice and said, "He did not wonder I was in the asylum--that my folks could not live with me at home, so they had to bring me to the asylum."

I admitted all his slang to be true, and said, "Yes, yes, doctor, that's so--you and I are here for the same reason, our folks could not live with us at home, so they sent us here." This roused the lion--and he could roar terribly when roused--but I said no more, and as my reply got the laugh of the table on him, he cooled off, but he never tried me on again. Whether he thought he had caught a tartar, or whether he thought I was a fool and not worth minding, he did not inform me; but one thing I do know, that ever after this he treated me with respect, and died in the asylum in about two years from that time.

As I have already noticed, this is called the first hall on the gentlemen's side, and is on the first floor above the basement. Between this, and what was then called the fourth hall, now I believe called the second, is a billiard room. The patients amuse themselves at this game, and some of them are expert players. I never took any interest in it; I never even took the stick in my hand to strike a ball while there,--neither did I ever elsewhere.

Chess, checkers, backgammon and dominoes, were the principal games played in the asylum, but in none of these did I take any interest; indeed, I never learned to play them. I think if all these games could be confined to lunatic asylums it would be just as well for the world.

As the time of retirement on this floor is just half-past eight in the evening, there is considerable time during the long nights of winter for some kind of exercise between dark and bed-time. So after the hall is lighted up, the patients betake themselves to such kinds of recreation as suits them best--some to reading, some to walking the hall in pairs, which is a good exercise, others engage in the different games practiced on the hall, while some will always sit looking blank, as though all the world besides them were asleep or dead.

There is a state of mind in that institution which I have thought would not be moved if the house were on fire. I once saw it demonstrated on the fourth hall. A young man was brought on to that hall by his friends from the city of Utica, subject to epileptic fits; these fits had injured his mind very much; yet he was as harmless as a child, and a greater mistake never happened than to take a child to the lunatic asylum who has fits, thinking they can be benefited by it; but if it is done to get the child out of their sight, and to throw the care of them on to other hands, why then, that alters the case; but if that were my object, I certainly should not send them there; I would sooner send them to the county poor-house.

But to return to my story. He was sitting at the table; I think it was breakfast; we had all commenced eating; in a moment he fell backward chair and all, with a terrible groan, foaming at the mouth, and uttering most horrible groans; I started up as by instinct; my knife and fork dropped from my hands, and I was about to take hold of him to take him up, when two attendants took him up and carried him to his room. But I observed that more than half of those at the table never looked up nor stopped eating. I made up my mind that if they had been asked after breakfast, what happened at the breakfast table this morning, the would have said, "nothing that they knew of." That was the state of mind I wanted to be in when I entered the asylum; then I should have had no trouble by anything I saw or heard. I do not wish the reader to understand that I now wish that to have been my state of mind.

Though it may seem a digression from the subject designed in this chapter, yet, while I am on the subject of epileptic fits, I wish to relate a fact which has come under my observation within a few weeks past. I was in a town in the northern part of Saratoga county about the 25th of June, 1868; and while there I was told that one of their neighbors was about to take their son to the Asylum at Utica, who was subject to epileptic fits, and they asked my advice. The brother of the young man who had fits was present. They did not know, that I am aware of, that I had any knowledge of the asylum. I asked them why they were going to take him to the asylum? I saw that he hesitated to answer. He finally said they thought it would be best. I asked him if they thought the doctors there could help one with fits better than other doctors?

I then told him just what I thought, "that many had been deceived by supposing they could cure epileptic fits at the asylum, and that they would miss it if they took him there for that purpose." They were entire strangers to everything pertaining to the asylum, yet I saw they were intent upon taking the young man there. They started with him the next morning, and took him to the asylum.

After the young man, the brother of the patient, had left, the family where I stopped explained to me the probable reasons why they were going to send him to the asylum. The young man had become of age, and was not capable of supporting himself; they were afraid he was getting, or would get, suicidal; he was getting to be a burden to the family.

His own mother was dead, and he had a step-mother. If they put him in the asylum they would get rid of the trouble of looking after him, and would save his support by throwing him into the asylum as a county charge. Yet they were not poor. There can be no doubt but many are sent there for similar reasons.

I was greatly surprised to find children in the asylum not more than six or seven years old. I saw two little boys there, one from Rondout, of about that age. Poor things, how I pitied them. They were very sprightly little fellows, but it was said they had epileptic fits. I would think it much more appropriate to send a child there that had the measles, or one troubled with worms, than to send one there troubled with fits, for, I think likely, they might cure the measles, and I am sure they would give medicine enough to kill the worms.

I will now return to the first hall, and give a description of the patients on that floor. I have already said that generally there are about forty patients on this hall, perhaps a little less; there are constant changes on this hall of patients. When patients are first brought in they are seldom left on this floor, though some are. Some come on this hall and never go to any other. They come and stay from three months to a year, and sometimes longer, as the case may be, and leave almost entirely ignorant of the general state of things in the asylum. And some of this class are well pleased with the asylum; this depends much on who they are, and what the state of their body and mind was in while there.

But the great majority first have a schooling on other halls, and, if very insane, are quite likely to be sent on the old eleventh, which, I believe, is now changed to some other number, but the hall is the same. This hall is the most like hell, in my judgment, so far as we have any knowledge of what hell is, of any other place on earth. I recollect when I was there, I used to fear and tremble, lest I should be sent on to the eleventh; and it was a common thing for attendants to scare patients, by telling them they would report them to the doctor and have them sent on to the eleventh hall.

This is a low small hall, on the ground floor, in the west end of the wing, made of brick, and, I think, but one story high. Here men are bound in fetters and laid in irons! Many of them are so crazy they are obliged to be kept bound, some in cribs, some hand-cuffed, some tied down in seats, some with muffs, and many of them in strait jackets. I am not censuring anybody for this, unless it be the patients themselves, who have brought themselves to this state by imprudence and debauchery. As to the treatment of these, I have no knowledge, only by hearsay. I have often heard many hard stories concerning their treatment, but there can be no doubt, that means that would seem to be rash has to be sometimes used, to bring to bearings some of these raving maniacs.

Their food, I understand, is as good as in any other department in the institution, but the manner of eating it is different. They are not allowed knives and forks, but eat with spoons; their food being prepared and put on their plates by the attendants. As these patients improve, they are changed to other halls more appropriate to their state, until some of them finally get to the first hall. Not that all come on to the first hall who get well and are discharged from the asylum, yet many come on the first hall from other floors, and many are discharged from the first hall and go home, making constant changes on that floor. I visited the asylum in April last, and found eight persons on the first floor, who were there three years ago, when I left; some of these seem to be fixtures. I could give the names of these, but perhaps they would regard it a freedom which I had no right to take; so I will forbear. There is a kind of mystery attached to the history of some of these men. One of them, a well, hearty man of about forty, who has been there about six years, and he told me he had not taken a dose of medicine since he entered the institution; and no man would think of charging him with insanity; and I have often said, and say now, that a man must be made of stern stuff that can remain shut up there for six years, in the prime of life, amidst the howlings and babblings of five hundred maniacs, and not become insane!

There are a number of others who have been there from ten to fifteen years, who show but slight marks of insanity; if any, perfectly harmless. There are many out in the world doing business quite as crazy as they. These, I know, are all groaning to get free, but their friends prefer to keep them there, and as property doubtless has much to do with the matter, they will be likely to die in the institution. Were they county patients, they would long before now have been set at liberty.

There is one case in the asylum that I will venture to name, because I am quite sure he would prefer to have me do so. This case is Frank Jones. He has been there a number of years. He is troubled with epileptic fits; these fits have somewhat impaired his mind. He is as harmless as a child; is capable of doing considerable business of a certain kind. He has his liberty to go about where he pleases; he does a great many chores for the doctor; goes into the city daily, to the post-office and stores; dresses very neatly, is perfectly honest and truthful, and can be trusted in any matter. He occupies the first hall; is a private patient, and is the son of Mrs. Jones, the owner of the Clarendon House at Saratoga Springs--one of the best houses in that place.

I have often had talk with Frank on the subject of his being in the asylum; he seems to feel bad at times that his mother chooses to keep him in the asylum. I had the opportunity of watching him for two years. I have seen him have his fits; he is very little trouble to any one when he has them; he generally so manages as to have them in his room on his bed.

It is true that this is not my business, but were he my son, and knowing what I know of asylum life, I should remove him to some private family, where he could enjoy the comfort of social life, if I did not want the trouble of looking after him myself; it could not cost any more than to keep him in the asylum.

There is one more man in the institution of which I will say a word in this connection. His room is on the old fourth hall, now called the second; this is Esq. Bebee. He was in the old asylum at Hudson, I am told, before the one at Utica was established; and on the opening of the one in Utica, in 1842, he was removed there, and I think has been there ever since. He was a lawyer of superior talents. I understood he fell from a horse and fractured his skull, that a portion of his brains ran out, and they were preserved.

He is a very eccentric man, and has a very lofty bearing. I have heard him speak a number of times, and have heard him make some of the most able and thrilling speeches I have ever heard from any man. He keeps his room the most of the time; has his liberty; goes where he pleases, but will doubtless die in the institution. He frequently shows marks of insanity, not by any low or foolish expression, but by some sudden outburst of eloquence, or some ludicrous and eccentric act.

He is always very tenacious about having any one come into his room. I once saw a poor fellow who hardly knew what he was doing, step into Bebee's room just as he was coming out. Bebee met him at the door, and with a lofty swagger, exclaimed, with a good deal of energy, "Scoundrel, many a man has been shot for a less offence than that." The poor fellow sneaked off without saying a word. One day he went to the city, I was told, and while out lost his brains, which he had always carried carefully done up in his pocket. On his return he said, "I have lost my brains out of my pocket--the people now won't believe that I have any brains, as I can no longer show them."

I recollect that during the time I was on that hall, Bebee went out on a visit to see his friends, and was gone some three weeks. It has always been a mystery to me why he should stay there. There is no doubt but he would have been discharged long ago, had he been a county patient.

I will venture to name another particular case with which I deeply sympathize, trusting that he will not be offended that I have made mention of his name. This is Alexander Hamilton Malcrum, a grandson of old Gen. Schuyler, and nephew of the celebrated Alexander Hamilton. He has been in the asylum quite a number of years--is a man of good education, having been educated at Hamilton College, and is not insane. It is true he is a little eccentric, and so are many other men out of the asylum. He is groaning to be set free--is capable of doing business--is middle aged. I regard it a great cruelty that he is kept there so long. I have had long and frequent talks with him on the subject. He has property. I think his brother at Oswego would interest himself to get him away could he know the real facts as to asylum life.

CHAPTER V.

The holidays of 1863 came, and I saw that the attendants, and many of the patients of the first floor, were busily engaged in dressing up the hall, the billiard room and the chapel, with evergreens. The chapel is in the fourth story of the center building, and is reached by three long flights of stairs from the lower floors, rendering it very hard for old and infirm people to reach it. At times I found it very difficult, on account of lameness, to ascend these stairs.

Above the chapel, in the fifth story, is a theater; this was fitted up the first year I was in the institution. This I suppose was done for the amusement of the patients, and during my stay there quite a number of scenes were acted, on the merits of which I am not able to give any opinion, as I am not acquainted with theatrical performances, having never attended one before I went to an insane asylum. I made up my mind, however, from what I saw, that they were very appropriate to a lunatic asylum, and that it is quite likely that in the first instance they were got up for the sole purpose of cheering and amusing disordered minds, and that by some unaccountable means they made their escape from the lunatic asylum, and have ever since been running at large through the world.

I think it would be one of the most humane and charitable acts that our country could perform, to pass an act to place all the theaters back into insane asylums, where they appropriately belong.

The first performance of the kind I ever saw, I think, was in January of 1864. The supervisor, Mr. Butler, said to me I must prepare myself to go down to Mechanics' Hall, in the city, as some performances were to be acted there that afternoon and evening. I begged to be excused, but there was no use in talking; so I got ready. I recollect that about a score of us poor lunatics, were marched off to the city. I shall never forget how I felt when I reached there. It seemed to me that all eyes were turned upon us, as they knew we came from the asylum; perhaps I was a little too sensitive on this point. I looked on, or pretended to look on, but I did it mostly with my eyes shut. I took no interest in the whole matter. I only went to obey orders; but I was a good deal like the horse who would not drink after he was led to the water. If there is any sanitive power in knowing we must obey, then I suppose I was benefited; so I walked down the hill, and walked up again. So we were a privileged people; we could go to the theatre, dance, play at billiards, attend church, drink whiskey or porter, and all sanctioned by law.

On this hall there is something of a library, containing, perhaps, five or six hundred volumes, besides papers, both daily and weekly, that are brought on to the hall; so that all who desire reading can have it. Patients from other halls frequently come down and get books, read and return them.

As to religious service, it is regular once a week, every Sabbath evening, so that all who desire to attend church can have the privilege in the chapel. Besides this, there are quite a number, such as the doctor pleases to select, who have the privilege of going into the city to church, accompanied by an attendant, who goes to see that they keep orderly and return home at the close of service.

I observed that people of all creeds were in the institution--Episcopalians, Methodists, Baptists, Presbyterians, Roman Catholics, Quakers, Unitarians, Universalists and Swedenburghers, so that no one denomination can boast that their members are never insane. I judge, however, that there are more Roman Catholics, compared with their numbers, than any others in the asylum.

Another inquiry arose in my mind while in the asylum, viz.: What class of the inhabitants form the majority in that institution? This is rather a hard question; yet, perhaps, we can arrive at something near the truth on the subject. In doing this I will divide society into four classes, as follows: first, professional men and men of study; secondly, business men, who have much care on their minds; thirdly, the common laboring class, which compose the great majority of mankind; fourthly, that floating unsettled class of men, who live as they can catch it, with no settled business, and indulge in drinking and in other loose habits of life.

Some of each and all of these classes are found in the asylum--doctors, lawyers and ministers of the gospel, with students from colleges, are often found in the asylum; yet their number appears small compared to the other classes, but it must be remembered that this is by far the smallest of the four classes of community. I judge that this class is as about one to fifty of the second and fourth class, and as about one to two hundred of the third or laboring class, so that if ten professional men are found in the asylum, with three hundred of the other classes, it would show a large proportion of professional men in the asylum. But I do not think that this class will average more than six to three hundred of the other classes; this is giving a very large proportion of professional men to the asylum compared to other classes. I think, perhaps, I have the numbers too high. All I can say of this class then is, that education and study is not a safe-guard to insanity, but sometimes may produce it; yet it is thought very strange by some, that a man of mind, study and education, should ever become insane.

There are some men who need never fear of becoming insane--their minds are not sufficiently active--they will never rack their minds with study--in a word, they have not brains enough to become insane. As to the second class, they are quite liable to overtax the mind with the burthen of their business. I judge this from their numbers found in the asylum. I cannot say, however, that I have seen as many of this class in the asylum, according to their numbers, as I have of the first class.

Of the third, or laboring class of community, there are a great number in the asylum. Many of these suffer in various ways, and from various causes. Some, by overwork, undermine their constitutions; some, by exposure to all weathers, become prostrated, and their nerves unstrung. And many in this class, as well as in others, have greatly injured their nervous system by the excessive use of coffee, tea and tobacco. It is a remarkable fact that but few men are found in the asylum who are not users of tobacco; and the universal cry of the patients through all the asylum for tobacco, is proof of this fact. I think there are five to one of this class in the asylum compared to the other classes; yet, perhaps, they number ten to one of all the other classes.

The fourth class is that reckless and unsettled portion of community that never look beyond present gratification, whatever it may cost. Rum, tobacco and idleness, constitute their chief study; habits unfixed; system in living never enters their thoughts; and though this is not the larger class of community, I doubt not but two to one of this class are found in the asylum to any other class of society.

It is a given fact that a great number in the asylum were brought there by their dissipation. It is not strange that many of all the classes mentioned should be found in the asylum, but to see the imbecile and driveling idiot thrown into a lunatic asylum, carries prima facie evidence with it, that the object in placing them there was not to prevent their doing injury to themselves or others, nor for their recovery from their unfortunate state, for many of these were born so. If the parents or guardians of these unfortunate cases are not able to support and take care of them, let these turn them over to the county where they belong, for it would be much better for such to be in the county house, than to be shut up in a lunatic asylum.

There is a striking fact that will appear to any observer who will take the trouble to read the printed statistics of the number of patients in the asylum at Utica, and the counties to which they belong. He will find that some of the remote counties send one, some two, and some none, while those near by will send scores. I presume that the large cities of the State, such as New York, Brooklyn, Albany, Troy, Buffalo and Rochester, that all these cities do not send as many patients to the asylum as is sent by the little city of Utica, which does not contain over 25,000 inhabitants! This may seem a startling assertion, but I have known at one time in the asylum sixty patients from the city of Utica.

Can it be proved that the above named cities ever had sixty

patients in that asylum at any one time? It would take a hundred such asylums to take all that the State of New York would furnish if each county should send as many as the city of Utica, according to their number of inhabitants.

Perhaps it will be said that this fact is all in favor of the institution; that Utica knows better the worth of the institution than places more remote, and this is the reason why so many more are furnished from Utica. I am fully satisfied that the citizens of Utica know no more about the private workings of that institution than the inhabitants of Clinton and Essex counties; and living near by renders them more liable to be deceived, and in the following manner: It is known by all the inhabitants of that region of country round about Utica, that the asylum is open every day at certain hours, for the reception of visitors. It is also understood by the managers and attendants at the asylum, that visitors are expected every day, more or less; so that all things are put in order before visitors come; every unsightly thing is put out of the way; all is still and clean as a ladies' parlor on the first halls, on both sides of the house; the time comes; the usher is at the door; the visitors are led through the first halls, look at the pictures and leave. What do they know by this running visit about the asylum? It is true, they have seen the neatness and order of the two lower halls--the lovely flower garden--the beautiful lawn spread out from the out-stretched and towering walls of the asylum, to the archway that leads to the street below; the view is lovely.

My daughter visited me in my prison-house after I had been there ten months, and she is a lover of the beautiful--she exclaimed, after she had feasted her eyes on all around in full bloom in the month of July, "O pa, it is a paradise; I should like to live here." Tears filled my eyes, though I had not shed one tear for a year; my grief had been too deep for tears. "Poor child," thought I, "I hope you will never be undeceived by being placed here as a patient."

No, it is not because the people of Utica know better about the institution than others that they send so many there. It is true they know the managers of the institution, the steward, and Dr. Gray. But Dr. Gray himself does not know one-half that is done in that place of deception. If I thought he did, and tolerated it, I should have far less respect for him than I now have.

I know a gentlemen living not far from Utica, of prominence and standing in community--a man of wealth and large business--has held the highest office in his town for years, and had often visited the asylum, and walked through its halls, and had boasted of the value and utility of such an institution, and was proud that he had taken an interest in the erection of so magnificent a pile--who does not feel now as he then felt--and why? Why? for the very plain reason, that since that time he has been initiated into the secrets of the institution. This man is no other than D. J. Millard, Esq., of Oneida County.

He was, like myself, unfortunately thrown into that institution as a patient. I saw him the day he entered it. I saw he was a man of more than ordinary ability; he was one of those business men I have described in this chapter; I formed his acquaintance in the asylum; he was not insane, his health became poor; his business lay heavily upon his mind, and he partially sunk down under the burthen. Difficulties magnified in his mind beyond what were the real facts. But an insane asylum was not the place to cure him; it was the very worst place, in my judgment, that could have been chosen for the relief of his mind.

Encouragement and cheerful greetings was what he needed, instead of imprisonment and seclusion from his business and his family. But he lived in spite of all these opposing influences, and came out of his troubles a wiser, and no doubt, a better man, for his sufferings. Would he recommend a friend to place one of his family in that institution?

He is not a man who is carried away by low and petty prejudices--he sees things in a broad and philosophical light; he believes that such an institution could be, and should be, a blessing to the State and Nation, and that it would be, were it conducted as it should be; but as it is, and as it has been managed for a few years past, he regards it a curse to the land, and unless reformed, will one day fall by its own weight.

The light is already breaking in through its dark and massive walls, and when men can be placed over it who can feel for suffering humanity, instead of glorying in a little power over helpless invalids, and seeking how they can make the most profits at the expense of the sufferings of their fellow beings, then, and not till then, will the darkness and gloom, which has so long hung over that prison house of death, roll off, causing the tongue of the dumb to sing, and many a bleeding heart to rejoice.

As to the nationality of the inmates of that institution, it may almost be said, that they are from all nations, tongues and languages. There is the American of the Anglo-Saxon race, the Welshman, the Scotchman, the Irishman, the German, the Swede, the Frenchman, the North American Indian, the African and the Jew. I have thought it partook very largely of the Irish race--I think so still; so that that institution may be said to be the world in miniature.

There are the rich and the poor--the black and the white--the wise and the ignorant--the learned and the unlearned--the Devil and the Saint--the Christian and the infidel--the drunkard and the man of temperance--the libertine and the man of chastity--the thief and murderer--the man of honesty and kindness--the child and the man of gray hairs.

CHAPTER VI.

The winter of 1863 and 1864 had nearly worn away, and I heard nothing from any of my friends, nor had I seen but one person that I ever knew before I entered the asylum. What winter clothing I had left were worn nearly out; my vest was very ragged; my pants were quite thin; as yet I made out very well for coats, such as they were. I forgot to say, that besides the broadcloth coat that I had taken from me, I had also a farmer's satin coat taken about the same time, and another given in its stead. This I regarded as an insult, for the one given me was old, rusty, cut in an other form, and quite too small, my arms extending some distance beyond the ends of the sleeves. I could not help laughing when I put it on. I never wore it three days while there, I keep it yet as a curiosity.

On thinking over all these matters, and looking at my ludicrous plight, I felt that Heaven and earth had turned against me. I then thought that could I have had a friend to whom I could have spoken freely, to whom I could have poured out the feelings of my heart, I could have got relief, but I had no such friend. What I had already said about my loss of clothing had only caused a sharp rebuke; no one would hear my story and pretend to believe it; so I was dumb.

I had been ordered into the dining room to assist in doing up the work after meals. This was awkward business for me at first. I had never been in the habit of washing dishes, but I commenced my apprenticeship feeling quite indifferent whether I succeeded in learning the trade or not.

After preparing myself for the work, by laying off my coat, and putting on an apron, as the custom was, I could not help comparing my present condition with my former one, ten months before. To say that I felt humbled and even crushed, are no words to describe my feelings at that time.

Soon after this, when in this condition, dressed in my apron, with sleeves rolled up and dish-cloth in hand, I was called to step into the side hall; I did so, and who did I meet but an old friend and parishoner, now living in Albany--his name was Hoxsie. He was very neatly dressed, but I observed he looked sad as he looked upon me, in my shirt sleeves, apron on, pants ragged, and my vest worn all out. I was not glad to see him while in this plight, for he had never seen me before only in the capacity of a Pastor, decently dressed. I know I appeared very much embarrassed and eccentric when we met. I did not know what to say or do. Many things rushed upon my mind which I wanted to say to him, which I could not, for I knew we were watched by an attendant, and every word would be marked and reported. I knew he did not understand all this; and besides this, I knew he had a right to expect that what I did say would discover traits of insanity, for all are supposed to be insane who are in the asylum.

I recollect the first thing I said to him was to ask him about my family, whether they were well, and where they were? He could give no information about them. I told him this was a horrible place; that he could know nothing about it by such a visit. I asked him what the people were going to do with me. I saw he looked embarrassed; he did not take my meaning. I meant he should take the hint, that I wanted my friends to interest themselves in getting me away. I pointed him to my pants, and asked him if I could not have a new pair. I doubt whether I said anything about the loss of my clothes, as we were watched. He made me an indifferent reply when I spoke to him about the pants. I saw I should get no help from that direction.

He seemed to be in a hurry, so he rose and left. He is a fine and good man, and if he ever sees this, he will know more about my feelings at that time than he then knew.

It was not long after this before the doctor ordered the supervisor to take me to the city, and get me a suit of clothes. We went down, but I felt a great reluctance in going; not that I did not need the clothes, but I felt somehow that I did not want any new clothes got me while there. I wanted to get away, and I feared if I got a new suit that I should stay until they were worn out, and my fears were realized.

It was left with me to choose such a suit as I pleased. I selected a strong, common suit instead of a fine one; in this I was right, as I stayed there until it was about worn out. I now appeared to a little better advantage on the hall. The patients are expected to dress a little better on this hall than they are required to do on the back halls.

April now came, and quite a number of patients, who had been on the hall through the winter, now left for home. I had made the acquaintance of these, and to some of them was warmly attached;

when they left my spirits sunk down for a season. I was left behind, and some of those who left had come into the asylum subsequently to my entering it. There was one Dr. Brown, from New York city, who left; he was a Quaker; a fine fellow, but subject to depression, having had some trouble, perhaps, of a domestic character. I was surprised a year after to see him through the window in the yard with raving maniacs. He saw me and hailed me. He had been in the asylum a number of weeks at this time; he soon came on to the first hall, stayed a few weeks, and left for home for the second time, long before I left.

One fact was quite observable in relation to patients, to illustrate: A man or woman comes into the institution a raving maniac, hand-cuffed, and hair dishevelled, foaming at the mouth and uttering hideous yells; they are ordered on the old eleventh, for instance. Nothing more is heard from them perhaps for three or six months, when all at once, they are introduced by the doctor, or the supervisor, to the patients on the first floor; they are sober and in their right mind; they stay a few weeks longer and return home.

But these, I find, are liable to a relapse, and often return the second, and third, and even the fourth time. These are excitable temperaments, and when their nerves become unstrung, there is no holding them, so they are brought to the asylum. They only want rest, and to be kept clear from excitement; any other place would be as good as the asylum if they could be controlled. Another comes into the asylum gloomy and sober, with his head down; is still and harmless; talks to none; shows no marks of insanity; except, perhaps, you hear him groan or sigh occasionally; he sets down alone. He stops perhaps on the first hall when he first comes to the institution; stays there three months, and perhaps a year; when he is found to be no better, but worse, he is finally placed on some of the back halls; gets no better, is changed from one hall to another, till finally is pronounced demented; he lingers on, and either become a fixture in the asylum or dies there! Such is asylum life.

I know one man, a dentist, who has been in the asylum ten times at least. There is a young man by the name of Bouck, from Schoharie, who has often been in the asylum. He comes under a high pressure of excitement; stays a few months and leaves. But while there, is regarded the lion of the establishment; fears nothing; is a giant in strength; will dash out windows with iron grates as though they were made of cobwebs; will climb on the side of the wall where no sane man would dare to venture. For such a case, perhaps the asylum is of some value.

As the spring of 1864 had now opened, I looked out with surprise that I had lived through the winter. I confess that when winter set in, I did not expect to see the leaves put forth again. Not that I was sick, but I did not believe that I could bear up under the pressure that lay upon my mind. There was some cause for this. A little before I left the fourth hall in December, I had a weak turn; I would attempt to rise in the morning after a sleepless night, and would fall back faint and weak upon my bed. Had I been anywhere but in a lunatic asylum, I should have lain down quietly until my strength had rallied, but I dared not do it, (I confess I feared the attendants' ire), so I would rally all my energies and get up, dress me and make my bed the best I could, concealing my weakness from the attendants, for I knew to make it known would not help me.

One morning when I was making my bed, the attendant stepped into my room. I then took occasion to tell him my feelings, and said that I did not know but I should be unable to rise in the morning with the rest, and if it should happen I wished him to treat me as favorably as he could. He replied that all the treatment I should get in that case, would be that he would wait for me just ten minutes after the signal was given for getting up. I replied that I should do the very best I could, and then must suffer the consequences. But by the blessing of God, I was ever after that able to get up, dress and make my bed, while I remained in the institution.

The impression was indelibly fixed on my mind, that for me to become helpless in that institution, would be the same to me as death. I was, soon after this, removed to the first hall. As the spring opened I went out with the men to work on the lawn. The first work I did out door was to rake the old dead grass off the lawn into heaps. It was then drawn off with hand-carts. I had had a broken arm the year before, which crippled my right hand so that I was not able to do much; besides this, I had not been used to work since I was a young man, and to be ordered about by an ignorant attendant boy, did not go down very smoothly; however, I tried to make the best of it. I suppose the main reason why I did not leave the institution without liberty was, that I knew the authorities had power to take me back without a new order, and hold me until legally discharged or released by the doctor.

Summer came, and I went into the field with other patients to work; the weather was hot. I recollect of looking about me and seeing a motley group of lunatics, some cursing, some yelling, while others were keeping up a constant ribaldry of blackguarding and obscene language.

I though of home and of friends; I compared my present state with the past; I could hardly believe this was a reality. I thought I would have given a world, if I had it, to have impressed on the minds of my friends at home, and the doctors there, my thoughts and feelings.

I thought of the convicts of a State prison that I had seen in the fields at work, guarded by attendants, as we were, some with chain and ball attached to their ankles. The only real difference I could see between us was, that they were not insane, and they were there for a definite period of time, and could look forward to that point with a certainty of being liberated, if they lived until that time; we were there to stay until doomsday, for ought we knew.

I recollect coming in from the field one day at noon. I was called to the supervisor's room; he took down a bottle and poured out a table spoonful of some kind of liquid, as white as water, and ordered me to drink it. I had learned before this, to ask no questions when anything was given me to drink. I drank it down; he repeated the dose, and I took it. He saw that I writhed under it. He said I must come to his room three times a day before eating, and take two table-spoonfuls of the contents of this bottle, until it was all taken up.

It was a large case bottle, holding, perhaps, a little more than a quart. I judged it to be the decoction of quassia wood; at all events, I had never taken anything before that compared with it for bitterness. Said nothing, but a strange feeling came over me. I was taking other medicine, as usual, besides this. I felt for the moment that they, seeing that I was doing well and gaining my flesh, took this course to kill me, by over-dosing me with medicine. It seemed to me that I could never live to take all that with my other doses, but I did take it and live.

But I did not believe then, neither do I now, that the doctors thought I needed this in addition to the beer and the other medicine I was taking three times a day. I have always believed it was given me to see if I would not resist. I had never once resisted taking anything offered, and never meant to, live or die, for I knew it would be forced down me if I did; I had frequently seen the operation performed on others, and I did not covet the luxury.

Perhaps this conclusion of mine will be regarded by many as unjust and unreasonable, who are unacquainted with matters in that institution, and of course will be laughed at by those who ordered the medicine. I would laugh at it too if I were they; it is the best way for them to dispose of the matter. Yet my opinion will be the same; I have my reasons for this. If I had been running down in health and appetite, confined to my room or to my bed, such a course might have seemed justifiable, but I was well, eating very heartily, working in the field every day with others.

July came, and I had heard nothing from my friends, and nothing had been said to me about writing to them. I had once asked the privilege of writing to the man that took me there, but had been denied.

I was sitting in the reading room one Sabbath afternoon in July; my anguish of mind was very intense, as I was considering my condition--that my present life was worse than a blank, shut out as I was from all knowledge of the outer world, and yet in a free country. I was not aware that I had forfeited my liberty by any crime, yet I was confined by bolts and bars, and if I was permitted to go outside, was guarded and watched by a set of ignorant, unprincipled hirelings. Such were my meditations, when all of a sudden the newsboy announced to me that my daughter had come and wanted to see me.

I was paralyzed--I could hardly believe it--I thought it must be some one else, for I knew she lived a thousand miles off. I rose without speaking and left the hall and went to the sitting room in the center, and lo it was my daughter. I shall never forget that meeting.

When she left us more than a year before, for the far west, I was in good health, and all was prosperity with us, and I was a man in the world like other men, and a father that she was not ashamed to own. Now we meet in a lunatic asylum.

I shall never forget my first words to her, even before I had enquired after the family, putting my face to hers, and pressing her to my bosom, I said, in a whisper, for we were watched--"for Gods sake never send one of the family to this place what ever the consequences may be." I doubted whether she took in my full meaning at first, from the reply she made, but afterwards I explained to her what I meant.

I have never doubted, but this visit was the means of prolonging my life, and of my final release from that prison. She remained two or three days in the city and visited me daily while she stayed.

I was permitted to walk out with her in the garden and through the grounds, I learned from her that the rest of the family were all in good health. This was a great relief to me. I told her many things, and explained to her the workings of the institution, as far as I thought it advisible.

I pledged her to keep me advised of all matters at home, and if possible to get me out of this place. I knew, however, that if she did write me, that all would depend upon the will of the doctor whether I ever received her letters. It is not very pleasant to know that a third person has the power to intercept all letters received from, or sent to friends.

She talked with Dr. Gray, and he made her believe it was best for me to remain in the asylum. I was permitted to visit with her in the city, and when she was about to leave, I applied to Dr. Gray to let me leave the institution, and go home with her. He was very decided,--and said, "as a state officer, he could not let me go." My heart sunk down.

The time came for her departure; I went to the city with her; she had her little boy, her only child with her, of nine or ten years old. When the moment came for separation, she and my only grandchild, to go to her mother, and I to go back to the asylum, my heart nearly died within me. I bade her and the child good by, and gave them my blessing. But, O God! What a moment was that to me, as I gazed after my two only children as long as I could catch a glimpse of them! and then said to myself, "shall I ever see them again?" None but a father can know how I felt at that moment. Ah, none but a father in like circumstances can know how I felt! An ordinary parting of parents and children is touching; but one of this kind is beyond description.

If a man is insane, no such thing moves him; he can see his children go and come unmoved and unaffected, he can see his children die and not be moved, all things are alike to him.

I returned to the asylum with a heavy heart, yet comforted that I had seen my only beloved children, and thanked God for the opportunity. By the coming of my daughter, I formed an acquaintance with some friends in Utica, who called occasionally at the asylum to see me.

I have passed over a circumstance which I will notice in this place. While on the fourth hall, in about the month of November, I observed a thick, stout built man brought in from one of the back halls, and introduced to the attendant. He had come from the eleventh hall; he was a bold and naturally a good feeling man, and, I perceived, a man of strong impulses; and of some cultivation, he attracted my attention, and I perceived he was highly gratified with his change. On further acquaintance I found he was a preacher from the New York Conference; his name I shall withhold.

He had been thrown into the asylum by his friends, I learned, in consequence of the high state of excitement his mind got into by over-working and much care. He was first put on one of the back halls, and soon got on to the eleventh. There they have rough work sometimes. He was under a high excitement when carried there.

Mr. Vallerly, the attendant, a strong Irishman, and not overstocked with patience, took charge of him. The Reverend gentleman supposed he understood his own business, and, therefore, was not very prompt in obeying the strict and iron rules of the attendant, upon which the Hibernian drew his fist and knocked him to the floor, in the meantime giving him a terrible black eye, which he brought on to the fourth hall with him.

This Vallerly had the name of being a perfect gladiator, and this, I suppose, is the reason why he was placed on the eleventh as an attendant. I ever after that was afraid of Vallerly. This hall is greatly dreaded by the patients; is regarded as a whipping post. I confess I always had fears of being put there. This Reverend gentleman expressed his high gratification in being removed from the eleventh hall, saying he felt raised at least fifty per cent.

He was free to talk of his being a minister of the Gospel; he observed that I said but little about my being a preacher; I told him I did not care to say much about it while in the asylum, not that I was ashamed of the Gospel of Christ, but thought it was a disgrace to the ministry to have one of its members thrown into a lunatic asylum. So deeply did this matter affect me that the prefix Rev. to my name on some of my clothing annoyed me very much. Was this one mark of my insanity?

There were two or three things which used to cause me to suspect sometimes that my mind was not right, that I was a little insane--yet these things haunted me more or less for the most of the time I was there. One of these was the fear of being put on to the eleventh, or some one of the back halls--another was that I should never get away alive, and that the life of a patient was counted of no value in the asylum, especially by some of the attendants, and that many were put out of the way here; that no one out of the institution knew, or ever would know how they came to their end.

Now I confess that if these things are proof that I was insane, I shall have to bear the charge, for I could not help coming to these conclusions from what I had seen and heard. And if these things are proof that I was insane then, they are proof that I am insane now, for so far as the two last things noticed are concerned, my mind has not been changed, viz.: that the life of a patient in that institution is counted of no value, and that many pass away from that place, that the manner of their coming to their end will never be known in this world by the people out of that institution.

This may seem like a most reckless and slanderous charge; but when it is confirmed by testimony that cannot be reasonably disputed, that unprincipled attendants, have frequently knocked down feeble and insane patients, kicked them unmercifully--dragged them by force to the bath room, when weak and feeble, plunged them into a cold bath, and scrubbed them with a broom-corn broom, throwing on soft-soap which would come in contact with raw flesh caused by blistering or other sores--trying to hold their heads under water to punish them for struggling against such harsh treatment.

Besides this, choking patients until black in the face in forcing medicine down them--locking up patients, however sick they may be, leaving them alone through the long night to shift for themselves the best way they can. If all these things can be proved to be true--and for myself I have not a doubt but they can be, will it be urged that the lives of patients in that institution are valued as they are elsewhere?

If such treatment as this can be proved to be true, is it difficult to come to the conclusion that many under such treatment sink down and die? That some patients are treated with great care and tenderness, is not doubted. The circumstances of the man makes all the difference in the world. Acts of violence and cruelty have been related to me by those who were eye witnesses, that would compare well with the most cruel treatment in Andersonville prison. But these witnesses were patients, and because they were patients, their testimony will be disputed. It is true they were patients, but not insane at the time they told me these things, neither were they ever insane in a way to rob them of their reasoning powers; I have no reason to doubt their testimony.

I will here give a few instances as the facts have been related to me, and the reader must judge whether they are true or false. A young man had been a patient in the asylum, and was, when I entered it, a young man of veracity and standing, the son of a clergyman; he lives not far from Utica. The name of the young man I shall withhold. He visited the asylum perhaps six months after he was discharged. He was now in good health, and was doing business.

While there he related to me the following circumstances which took place while he was in the institution as a patient, on some one of the upper halls; I do not recollect the number of the hall. He said: "There was a poor skeleton of a man on the hall as a patient, who did not weigh more than about seventy pounds; that this patient was ordered into the bath by the attendant; that he hesitated, and struggled to prevent going in; that the attendant called him to his aid; that he did help the attendant to put this poor creature into the bath; that some force had to be used."

And as I understood him the water was cold; "they there washed and scrubbed him as the custom was, that the man went into spasms and died in four hours." This young man said, "he was sorry he helped the attendant." Will this relation be said to be false?

Another case was: that a poor patient was ordered to do something; he did not instantly obey; he was thrown down by the attendant; he struggled and showed resistance, as the most of men would, and especially one insane; the attendant fell upon the breast of the patient with his knees and broke in his breast-bone, and he died!

While I was on the fourth hall, there was a man brought there as a patient, who they called Major Doolittle, a gentlemanly kind of a man; I became acquainted with him; he told me he was uncle to C. Doolittle, Esq., of Utica, a celebrated lawyer of that place. I left this man on the fourth when I went to the first hall. I observed he began to run down in health about the time I left the hall.

I could never discover that he was insane; I could never conceive why he should come to that place; I had a hint that property had something to do with it, as I heard he was rich, but of this I have no certain knowledge. He continued to run down slowly; he was an old man, and I observed was quite notional, not more than the most of old people generally are, however.

He became at length quite helpless, and the attendant had to assist him into the bath. There was an attendant on that hall at that time by the name of Smith, from North Carolina, as John Subert had now left. This Smith was as cruel as an Arab. I was told many things which he did; among the rest, he would throw Major Doolittle into the bath and scrub him with soft soap, until he would groan horribly, while Smith would laugh. Suffice it to say the major died in the asylum. I understood the cruelty of Smith was the cause of his dismissal soon after. I know he left the place, but as to the cause I know nothing, only by hearsay.

CHAPTER VII.

I brought down my narrative in the preceding chapter to about August, 1864. All things went on in about the same monotonous manner, taking medicine three times a day, eating three meals, working some in the field and walking out with the attendants.

When September came, Sabbath school celebrations, and picnics of various clubs were frequently held in groves near Utica. To these, some of the patients had frequent invitations to go. I was generally invited to go to these, and frequently went, but I cannot say that I enjoyed them, I could enjoy nothing of this kind while known as a lunatic, in a lunatic asylum.

Some from there seemed to enjoy themselves just as well as if they had been the superintendents of the schools. There is a state of mind that is not unfavorably affected by placing them in the asylum--such for instance as one under the influence of hallucinations--there are many in that institution who believe they own the asylum, they think they run the institution with all its machinery!

As I was there during the war of the rebellion, I found many were brought there through the war excitement; some believed they were brigadier-generals; others believed they had been in the war and in many battles, who never saw a battle-field, nor ever shouldered a musket.

On one occasion, I recollect that a large number of patients accompanied by the attendants were on the ground. While there, the patients were allowed to stroll around the grounds and mingle with the people about as they pleased, the attendants only taking care that they did not leave the inclosure. I saw it was a good opportunity for any who desired it to run away, and I had but little doubt but on our return home our numbers would be less than when we came, and so it turned out; this only shows that the great majority of the patients are held there against their will--this, however, is no objection to the institution itself.

Another incident I also recollect which happened on the ground at one of these celebrations--a poor boy the son of a widow climbed a tree, for the purpose of fastening a rope in its top for a swing for the children, his foot gave way and he fell to the ground, breaking one of his legs and receiving other very bad bruises. He was taken up of course, placed in a carriage and sent to his anxious mother, with a good contribution from the people to help repair the damage.

I did not run away, I felt some as Paul did when requested to leave the prison and would not. I recollect that in this instance I was rather one of the privileged. I rode to the ground in a buggy with the Supervisor, the distance being about two miles. This was the second time I had taken a meal outside of the institution since I entered it, and it seemed quite refreshing; more especially so, as we had such a dinner as is never got up in the institution. I remember of eating very heartily.

There is one thing which, perhaps, I should have noticed before, but it will come in quite as well at this stage of my narrative. About the third day after I entered the asylum I was sitting alone in a very melancholy state of mind, when I saw a man approaching me which I recognized as an old friend in 1848 and 1849, in Columbia county. I was shocked, I felt both glad and sorry to see him. I rose, he took me very cordially by the hand and said, "Brother Chase, how do you do?" I felt greatly embarrassed, choked up, turned either red or pale in the face, could not tell which, did not know what to say--I dare not say I was well, for I was in the asylum as a patient, and I did not feel sick, so I stammered out--"Col. Drier."

I had known him when under very different circumstances--I was the pastor of the church in that neighborhood, he had often heard me preach, he was also a Minister of the gospel, and now the steward of the asylum, and was at the time I met him on the hall. He said a few kind words to me, which I do not now recollect, neither do I recollect what I said to him, if I said anything.

I wish here to record, that Col. Drier, the steward of that institution, is a man, a christian, and a gentleman, always mild, always sincere, patient to hear all the requests of the patients, and though he could not gratify all their whims, he nevertheless so treated them, that all loved him, and as soon as he appears on one of the halls, the patients flock round him like hungry children round their mother. I never asked him for a thing that he denied me. I never heard of his doing a low or a wrong act in connection with that institution.

The fall passed away, and I began to be restless that I had heard nothing from my family since my daughter left in July, except one letter soon after she left; this was from Sandy Hill, where her mother was living at that time. A letter soon came, however, from Illinois, stating that her mother was with her in that country.

On receiving this letter my mind was greatly relieved; the mother was now with her only child, and though widely separated, I felt perfectly easy regarding the welfare of my family; I was only in distress that I could not be with them.

How often did I think that could the doctors enter into my feelings for one hour, and make them their own, that I should soon be dismissed from the asylum. But I now made up my mind to never say anything more about leaving, as the doctor once told me that my own opinion would weigh nothing with them in relation to my own case. I saw that a patient was a blank in all matters of opinion.

It is the custom in the institution, when the doctor enters the hall, for the inspection of the patients, for the attendant to walk by his side; and unless the patient is an old fixture, and not accounted much insane, the doctor asks the attendant the questions he wants answered, instead of the patient. This is, no doubt, right in many cases, but to apply the rule as it is generally applied, great injustice is frequently done to the patient. The questions for instance are: "How does he sleep nights?" "What is his appetite?" "Does he talk much?" "What is the state of his bowels?" "Does he take his medicine regularly?" The patient stands by, makes no reply; the attendant answers all these questions. I have stood by and heard the attendant answer these questions in relation to my own case. "Does he sleep well?" "Pretty well," is the reply of the ignoramus, looking blank at the same time, and why should he not look blank? What did he know of the patient during the last night? The patient was locked up in a room perhaps two hundred feet from the room of the attendant, and the attendant fast asleep, while, perhaps, the patient laid and rolled from side to side upon his couch, and never shut his eyes during the whole night. I have heard this answer concerning myself, "pretty well," when I knew I had not slept one wink; and so with about all the answers.

There are a good many little things in themselves, like this, that are very annoying to a mind that is not insane, and yet somewhat sensitive. Being always fearful that I might accidentally violate some rule and thereby fall under censure, I was always on my guard, and I can now recollect many things in which I was over particular.

One small affair I do not forget; it happened in the chapel on Sabbath evening. The second attendant took charge of the patients on the first hall that evening. He was an ignorant, self-conceited, over-bearing little Irishman. I took my seat in the chapel as usual, and had always supposed I knew how to behave in a church, as I had been a preacher forty years. I threw my arm upon the back of the seat, and as service had not yet commenced, cast my eye over my right shoulder; I had no particular object in view; he saw it, and thundered out, "Chase, turn yourself about, and sit up in your seat." All in the room heard, of course. I turned my head slowly around as though I did not hear, but I felt; yes, I felt that if it had been any where else than in a lunatic asylum, and he had said it, he would have wished it had not been him; but I never mentioned it to him afterwards; and as he and the first attendant soon after this had a falling out, he was discharged and went to Canada.

This first attendant of the first hall is also supervisor of a number of halls. His name is D. Pritchards, and a better man cannot be found for the place he occupies. I never saw him in a surly or wrong mood of mind, always cheerful, always kind, never over-bearing, never delighted in afflicting a patient; if he had any fault, it was that he was too fraid of afflicting or crossing a patient, or an attendant under him. The whole house like him on all the halls. I feel glad to give him this tribute of regard and respect, as he always treated me with a brotherly kindness, and did all for me that lay in his power. I find he occupies the same position in the institution still, and I hope the day may be distant when he shall leave.

My object in writing this sketch, is not to find fault or pick flaws with this institution, for there is no institution in the land of what ever kind or character, but has its enemies--this is all understood; but because this is so, it does not follow that an institution cannot become rotten, and that the people have no right to investigate its secret workings.

The winter came, the winter of 1864 and 1865, it was December, I had been occupying a small room by myself for the last three months, there were some reasons why I did not like it as well as some other rooms, yet I did not mention it as I liked it much better than the dormitory where I spent the winter.

Unexpectedly to me, the supervisor took me by the arm and led me to a very fine room in the center of the hall, the best room on that floor, having a fine clothespress and all other conveniences. He said to me I was to have that room. I could not see the point; I felt encouraged, for it seemed to me that they would not let me have that room long, so I somehow conjured up in my mind the notion that they meant to discharge me soon, and as another man wanted the room I had been occupying, they would give me this as it was not occupied, for a few days until I were discharged--this was a fine picture I drew in my mind, and one that suited me--little thinking at the time that this room was to be my home for just one year--which was the fact.

I put down my own carpet, had a good field bedstead and good rocking chair; a washstand, bowl and pitcher, which the rooms did not generally have--a good new bible was presented to me; a looking glass and a lock on my clothespress. I could not complain of my accommodations, and anywhere but in an asylum, I could have been quite happy.

As to the beds of the institution, no fault could be found with them. First, a straw tick, always kept well filled; next, a good mattress, three good cotton sheets and coverlids plenty, besides always next to the sheets, thick woollen blankets for winter; the outside one invariable a white counterpane; the pillows were not all of feathers; they were mostly of hair; mine, for the last year, were feathers.

About this time a tall, white-haired, well dressed man came on to the hall, acting very cheerful, and I saw all hailed him as an old acquaintance. He seemed to be perfectly at home. I soon learned, by his conversation, that he had come of his own accord alone; he had been there the year before as a patient, and having wintered well, and got quite fleshy, he left; but he thought the asylum would be a good place to winter in again. So he came back; put himself under the care of the doctor; gave him his check on the bank for nine hundred dollars as security for his keeping, and commenced operations under high encouragements.

It was not long before he began to complain that they would kill him with medicine; this was something he had not bargained for, as he was not sick but came to spend the winter in a quiet way with those he knew, as he had no family, his wife had died and he was left alone. He remonstrated against taking the medicine, but all was in vain. I told him it was "good enough for him, if after he had been there once and knew what he knew about the institution, to come here again of his own accord, was a mark of madness."

He would take the medicine, then swear, and curse the doctor for forcing him to take medicine which he did not need. He finally made up his mind that they meant to kill him with medicine, as they had got his money. It was most aggravating it is true, for the man needed no medicine, but either the medicine or the thought of it threw him into great agitation of mind, and not having a very strong mind he became nearly distracted.

Fearing that they meant to kill him by dosing him, he shut himself into his room, put his bed against the door, and barred it the best he could. The attendants found the next morning his door barred, and all fast, they of course burst it open, and such an outcry was never heard! He thought then of course he was a gone case. He roared and blubbered--but there was no use, he had to take the medicine.

He was now removed into the dormitory with other patients, in the same room. He finally concluded to take the matter into his own hands--he let me into the secret. It was to take the medicine in his mouth and walk carelessly away to his room or to the washroom and spit it out; he was very successful in this. I suppose for three months he did not swallow a table spoonful; yet it was given him three times a day. In the spring his son came and took him away: he went cursing the institution.

The asylum was now very full; some enlargements were made for patients. Some time in the fore part of this winter, as near as I can now judge, I saw a poor skeleton of a man come into the hall leaning on the arm of a man on one side, and on the other on the arm of a lady; he looked haggard, and I thought he was in the last stages of consumption.

They led him to one of the dormitories and placed him on a bed. I thought it strange that they would leave such a man on the first hall, as the sick and feeble were generally assigned to other apartments; I soon learned the cause of this. He was a merchant from West Port, Essex county, and a man of some means; his disease was dyspepsia.

He was advised to go to Utica asylum for a cure, as the doctors there were so very skillful. He thought it like any other hospital, that he could stay as long as he pleased, and if things did not work favorably he could leave when he pleased; and as his friends brought him there, and he paid his own bills, they wanted him left on the first floor.

His friends left, and he was left there weak and feeble as a child; I think I never saw before a man's limbs so very small. I pitied him; I knew he and his folks were sold, but I dared not tell him so. His appetite was very poor, and what he did eat distressed him, and he was in the habit of vomiting it up. He had a habit, when his food hurt him, of placing his head down lower than his body, which he thought helped him to vomit.

The doctor forbade his using any means to assist him to vomit. He was sly, and would vomit out of the window to prevent detection. He was soon after removed to another hall; and on passing through that hall a few days afterwards, I found him bound down to his seat with straps, to prevent his getting his head down. He looked wishfully at me. I pitied him, but dared not say anything to him; here he stayed for a long time.

At length he was brought back upon the first floor. His wife came to see him, but the doctor did not permit them to meet. He wrote to his family and read the letter to the doctor, representing things all right, but had a slip of paper prepared, counteracting what was in the letter; in this slip he begged them to come and take him away; this slip he put into the envelope with the letter.

His wife came, and demanded to see him; she did see him; she resolved to take him home, but the doctor remonstrated and she left him. This afflicted the man. He finally got some better, and walked out with me; for at this time I had my liberty to go out alone, when I pleased and where I pleased. He could not walk far at a time, but was anxious to walk out every day. At length he would stay out after I went in, sometimes for half an hour--he began now to lay his plans to run away, as the doctor would not give his consent to let him go. He one day stole the keys, and came very near effecting his escape, when he was detected. He did not deny the fact, but told them that he did it to get away; that he had done nothing to forfeit his liberty; that he was under no obligations to them; that he paid his own way.

His mind was now intent on leaving; he had written home for money, and it had been intercepted by the doctor; he resolved to go without money. He walked out with me as usual; he prepared himself by putting on all the clothes he could. I knew nothing of his plan; he lingered; I went in; he did not come in, and has never been in since. He went down to the depot by a back street, went to Troy, found friends there to help him on, and got home safe. I doubt whether he will ever go to Utica again to be cured of dyspepsia.

And though this man has a perfect dread of the asylum, there are men, however, who like the institution, and think it the best place in the world. It has been urged that those who so dislike the institution are those whose minds are not right; they are a little insane still, that if they were perfectly sane they would like it--that those who like it are sane men.

Let us see how this matter stands; they that like it are sane men, and those who dislike it are insane. I know a man who likes the institution, who has been in it as a patient for fifteen years; this man is known very widely, in Utica, in Hoosick and elsewhere; his name is Mosely. Is he a sane man? What does he say? He says: "It is the best institution on the globe, and that Dr. Gray and himself and his Bible, and the State of New York, the asylum, his farm in Hoosick, and his new house, are all one thing; that they all perfectly agree, and that it is the best institution on the globe." Now who can resist such an argument as this? Such are the kind of sane men who like the institution.

That there are men who work in and around the institution, and have for years, who see nothing very exceptionable in any of its departments, may be all true. So there may be men who are employed in and around State prisons, who see nothing very exceptionable in them. But this proves nothing at all.

There is a vast difference between skinning or being skinned. Let those who have been in and around the institution, and think they know all about it, let them go in as patients, let them go through all the degrees of initiation, until they get a diploma, then ask them whether they can recommend it to the world as the best institution on the globe?

As I wish to give credit for every good thing which happened during my stay in the asylum, and as I have passed over one thing, I wish, before I enter upon my last summer's history; to notice it.

The thing referred to, which was passed over in its proper connection, was our Thanksgiving dinners of 1863 and 1864. I was on the fourth hall at the dinner of 1863. I think it was about the 20th of November. I thought it a grand dinner; fifty turkeys were dressed, stuffed and cooked for that dinner for the patients.

I took a kind of philosophical view of it when it came on the table. The first thought was, after taking a glance at the whole thing, what a contrast. Now it must be understood that our common every day fare was a very stereotyped edition. It was bread and meat, and meat and bread, with a little butter, twice a day, and cheese, pickle, and pie, Sundays only; and I was always glad when Sunday came, for the sake of the pickle and cheese, though the cheese was a very small piece. I am not fault finding, only noticing the contrast. There was a fine roasted turkey on each end of the table, bread, butter, cheese, pickle, pie of the richest kind, roast beef; then came on nuts, confectioneries in abundance, with raisins and apples.

I think I must have been a little "luny" just at that time, for I confess I was so afraid that some of us would over-eat of this rich dinner, that ten to one if we did not have half a dozen deaths in less than eight and forty hours afterwards; for this dinner was not confined to one or two halls, but was general. At all events, I was so afraid of making myself sick, that I was foolishly reserved in eating; I ate scarcely any of the turkey, and, by the way, I never liked turkey; I ate no pie, I thought it was too rich; I made my dinner of stuffing, sauce, bread, butter, and confectioneries. I was not sick, and I heard of no deaths on that account.

The next Thanksgiving dinner, of 1864, was on the first hall. It did not make so deep an impression on my mind as the first, for it was not exactly like it, we had no turkey, nor butter for that dinner; but we did have a very good dinner, with a dessert and confectioneries to close up with.

Another spring now came, the spring of 1865; I had made up my mind to go no more out to work; I had got above work by this time, though I was better able to work now than I was the year before;

yet if I had been ordered out, I suppose I should have went, but very little would have been the work I should have done; as it was, however, I was not ordered out.

I had quit the dining room six months before this, except to eat my meals, as the Supervisor had told me I need not work there longer unless I pleased; so I quit it, and took to sweeping the hall for exercise every morning after breakfast. There were a number of men on the hall who were excessive eaters, but not one chore could be got out of them, except to make their beds and sweep their rooms.

The floor of the hall had to be scrubbed and washed every Monday morning; this gave us a little good exercise. The cleaning of house came on this spring, as usual; this is quite a business; the patients can have employment in this for a number of days. While this is going on no visitors are received. The windows are all taken out and washed, the mouldings and casings all scoured, the bedsteads all taken out of their rooms, the beds put into a pile and the bedsteads scoured and thoroughly saturated with kerosene, to prevent the vandals from eating up the patients. All the rooms are then whitewashed.

The bedsteads are prepared with strips of sheet-iron instead of cords to lay the beds on; this, perhaps, is an improvement.

One particular incident I cannot pass over without recording. Some time in the course of the fall or winter of 1864-5--I cannot be particular here as to the exact time--Dr. Gray came on the hall accompanied by a man in regimentals; a dark, curly black-haired man, rather slim, but carrying a decided look and apparently a firm will, and, as I inspected him from a distance, he looked to me, as though he could hew a man in pieces with all the sang froid of a Roman gladiator.

The doctor introduced him as Dr. Shantz, a surgeon from the army, and from this time was to be the attending physician on this side of the house. I had dreaded the one we had before, but now I thought we had got a Rehoboam, who declared "his little finger should be thicker than his father's loins; that whereas his father had chastised them with whips, he would chastise them with scorpions."

Such were the views I had of Dr. Shantz when I first saw him. He commenced his rounds of visitation, but I shunned him as far as I was able to do so, till some observed it, and thought I treated the doctor with great coldness. I was afraid of him.

At length we came in contact. I found he had a good mind, penetrating and scientific; I found he loved books, and was a good observer of nature, and withal was not an infidel; my fears fled. I soon found that he could not only reason, but was willing to hear others. After I had thoroughly weighed him in my own mind, I resolved on an experiment. For more than a year and a half I had now taken medicine three times a day, and was now, besides this, drinking strong beer before every meal, as to the medicine I had no doubt but it injured me, and I felt that I was like a candle burning at both ends, the pressure of the asylum on the one hand and the medicine on the other.

And so I contrived to evade taking it, by spitting it out. I confess I did this for more than three months, and I knew I felt the better for it. I will not stop now to argue the question of the right or wrong of my course, as I was not treated as a moral agent. I simply state facts as they were.

I told the doctor I would like to have an interview with him in my room if he would admit it. He said he would do so, and not long after this he came to my room and gave me a fair opportunity to tell him all that was in my heart.

I gave him a brief history of my coming into the asylum, the causes that led to it as far as I knew, what my feelings and state were before, and at the time I came there; how matters had gone on with me since I had been there; what my appetite was, my general state of health, and how I felt at that time; and closed by telling him that it appeared strange to me, that the manner of doctoring here should be different from the manner out of the institution.

In this particular I referred to the continuance of medicine of the same kind for a year or more, three times a day, without reference to the state of the patient. I told him that it appeared to me that when a man was well and appetite good, he did not need medicine; and finally begged him to take it all off.

The medicine was dropped off, and oh! how I rejoiced, not that I had swallowed it for the last three months, yet the idea that it was no longer offered me was a great relief. The bloating of my bowels and limbs ceased, and I felt much better. When it was no longer offered me, I felt like a new man, and hope sprang up in my mind. The beer was still continued; after a while I introduced this subject to the doctor.

I told him I felt quite well, and I could not see that I needed beer for my health, and begged him to take it off. He thought I was mistaken about its not benefitting me, but said he would take off the beer and substitute a little sherry, with an egg, three times a day.

I begged to be excused from taking the wine; so he took off the beer, and from that time until I left the institution, which was perhaps three or four months, I took nothing, and I know I felt the better for it.

So I found in Dr. Shantz a "man, a gentleman and a friend." I could not have been more kindly treated by an own brother than by him. When I left the institution, I felt that I had left behind a friend and a benefactor. I think him just the man for such an institution. I have had one very agreeable visit with him since I left the asylum.[C]

[C] Dr. Shantz said to me, at the time of this interview, that I ought never to have been sent to the asylum, and that if he had been one of the physicians who examined my case before I was sent there, he should not have admitted it.

He left the institution at Utica since I left there, went to Minnesota and founded an asylum in that State, of which he takes the charge. I understand he is doing well. If I am so unfortunate as to go to a lunatic asylum again, I beg my friends to take me to Minnesota and place me under the charge of Dr. Shantz, but never take me to Utica.

CHAPTER VIII.

I think it was in the spring of 1865 that I saw a man walking up the hall, who I recognized as an old friend from Fort Plain; we had been warm friends for a number of years; I had once been stationed there, as their pastor. Anticipating what he might think of my state of mind, I said to him the very first word, that I wanted him to look upon me as the same man that I used to be, and not to talk to me as though he thought I were insane.

The meeting was affecting to both of us. He took dinner with me. He had a talk with Dr. Gray about the propriety of my leaving the institution soon. I accompanied him to the depot when he left, and I am sure he was satisfied that I was not insane. Since I left the institution I have visited him twice, and once spent with him an agreeable Sabbath, in preaching in the same house that I did in 1850 and 1851.

August came, and the 23d of August came. Two years had now rolled round since I entered the asylum. I had said nothing about leaving since my daughter left, which was now more than a year before. My general impression had been, ever since I entered the asylum, that I should never leave it alive; but, for a month or two before the two years had elapsed, hope had begun to spring up in my mind; and when the two years were ended I hoped the doctor would tell me I could leave. But no such welcome message came, till at length, about the 23d of August, I said to the doctor that two years had now passed since I came there, and if I were ever to leave, I though the time had fully come. He replied very promptly, that "the two years had nothing to do with it; that when I got well I should go." "Get well!" I replied, "if I am to wait for that I do not know when it will be, for I did not know that I was sick." I then said: "Doctor, do you think I shall ever get away from here?" He answered, "Yes; there are some things we do know, and we know you will go away." I said, "Yes, I know, too, that I shall get away, either dead or alive; but how long, doctor, do you think it will be before I can go?" He answered, "Two or three months, if you get well, and your folks come after you." I said no more, but I stuck a pin down there. "Two or three months," I repeated to myself; it seemed short to think of.

I now felt that I had some grounds to hope; the time was limited to two or three months. Time now began to hang more lightly upon me. Mr. Harvey visited me during this fall, this is the man who was my attendant, when I went to the asylum. He observed that I was more like myself, that I appeared more life-like. And why should I not appear more like living? the medicine was taken off, the time was limited to three months, that I was to have my liberty once more and go where I pleased.

There was a man by the name of Fenton, a patient who used to accompany me in my rambles this fall, through the forests, fields and city. He was one of those eccentric, poetic, wiry, excitable creatures that would astonish you with his outbursts of wit and humor, making a very agreeable companion to help while away the gloomy hours spent in an asylum.

I left him in the menagerie, as he used to call the asylum, when I left; but I learnt that he soon after got away, and has written me two or three times since, sending me some of his poetic productions, which will compare well with our best American poetry.

As the time drew nigh for me to leave, the steward took me down to the city and told me to select just such a suit of clothes as I chose. I, of course, got me a good suit, with hat and boots. I was now prepared to leave, so far as I was concerned, with the exception of money.

When the three months were out, which brought it to the 20th of November, I reminded the Doctor of his promise. "Yes, yes," he said, "I have written to your son to come after you from Illinois;

when he comes you can go." I thought of the matter. I wrote to my folks to not come after me. I felt indignant that my friends should be required to come a thousand miles and spend a hundred dollars to accompany me back, when I knew I was just as capable of traveling alone as I ever was.

I sent to a friend for money. He sent me a draft. I told the Doctor my friends were not coming after me; that I was capable of traveling alone, and that I must start by the 20th of December, as I did not want to stay through another winter. Besides, I wanted to be with my friends during the holidays. He tried to prevail on me to stay another week.

I told him my clothes were packed in my trunk; that I had written to my friends that I should be there about that time, and I could see no reason why I should stay longer. I told the steward I must go to the city and get a few things before I started. I did so.

The last supper was now ended that I ever expected to eat in that house, as I was to start at eleven o'clock that evening for the west. At the close of the supper, a call was made for a speech from me before I left. The call was sudden and I was embarrassed, as I had not spoken in public in two years and six months. My own voice was strange to me.

I rose and addressed the company, about forty in number, they all seated at the table, with a few impromptu remarks. I referred to the length of time I had been there; that I had sat just two years in the same place at table, the changes that had taken place, the trials we had passed through, and encouraged them to hope on for their deliverance. I bade them all good-bye, with the best and most kindly feelings of sympathy, I trust, on both sides.

When the hour to depart arrived, the supervisor and house steward accompanied me to the depot, carrying with them a box of the choicest kinds of eatables for my accommodation on my journey. A sleeping-car was engaged, the signal was given to move; I shook my old keepers heartily by the hand, bidding them good-bye with unfeigned good feeling, and shot out of their sight--took my berth, and waked up in the morning in Buffalo.

I continued my journey through Ohio, around the lake to Chicago, and from thence on the great Central to the place of my destination, and found my family and friends in good health. But, oh, the change! To sit down in a private room by the side of a stove, with my own children, once more to eat with them at the table, to retire when I pleased without hearing that old stereotyped sound--"Bed time, gents;" to go out and in as I pleased, furnished grounds for the most profound gratitude to him, who had so mysteriously preserved me without harm through all my dangers and fears, and who had brought me safe to once more see my loved ones, and enjoy their society without fear of interruption.

And now, in the close, I have only to say, that, though it may be humiliating to spread abroad the knowledge that I have been an inmate of a lunatic asylum, yet, if by publishing this sketch, the people in general shall become better informed of the true character of asylum life, and thereby prevent the suffering of some poor, unfortunate victim to mental disease, I shall be amply compensated for all my humiliation.

* * * * *

P.S. Although my history is closed, yet there are a few things of a miscellaneous character, that I ought to notice, to make the narrative complete.

One is, that very much was said about the violation of the rules of the institution; patients were continually admonished not to violate the rules. I was very fearful that I might, by some mistake or oversight, violate the rules; I therefore sought to find out just what these rules were, that I might know the law. In doing this I became perplexed. I could find no code of laws or rules that were fixed, that could be possibly violated. I found a law, or rather a custom that amounted to law, which was fixed and unalterable, and there was no danger of this being violated. This was, that the patients must retire at just such an hour and rise at just such a time in the morning; that they must eat at just such a time, take medicine at just such a time, and no man would dare to violate these rules unless he loved punishment.

As to all other rules, I found them as variable as the circumstances of the patients were various. What one could do with impunity, I found was a violation of the rules by another. I was at first perplexed with this, yet the patients were constantly warned not to violate the rules. All the rules there were was the will and word of the doctor, who made rules and changed them just as he saw fit.

There was another thing held up very prominently to the patients, and also to outsiders. This was, that patients are not obliged to work either in the house or in the field, unless they chose to do so, and that no coercion is used either by the attendants or superintendents.

And this doctrine some believe; and indeed it is true with a qualification, but that qualification spoils it. The fact is simply this, that if a patient is told to do a thing whether it is to work in the house or in the field, that if that patient does it, all is well--if not, the patient must take the consequences, perhaps that patient is changed to some other hall, provided he or she is on the first, or some other lower hall; but suppose a patient is on the ninth or tenth or the old eleventh hall, and is told to do something and refuses?

Perhaps they would not be removed, for to remove them would be no punishment; but would the attendants on these halls submit to it? No one had better believe this. It is precisely in the asylum as it was in a certain school in this country; a boy was punished for violating the rules of the school, the teacher punished him; the boy made complaint to his father; his father told him he need not obey the rules of the school unless he chose to, but must go back to school.

The boy returned the next day and was punished again; he again made complaint to his father, the father still told him that he need not obey the teacher unless he chose to do so, but must return to school, he went the third day and was punished as before, he again made complaint to his father. His father then told him that he need not obey the rules of the school unless he chose to do so, if he preferred punishment, rather them to obey, but to school he must go.

By this time the boy waked up; he saw it was punishment or obedience; so with patients in the asylum, they are not obliged to work unless they choose to do so. But it is a base deception to pretend that patients are not obliged to work in the asylum.

I would recommend that all men who are sent to the asylum be permitted and advised to let their beards grow, and not shave at all during their stay there, especially on any other hall except the first, for the attendants do all the shaving; the patient is not permitted to shave himself, except on one or two of the halls, and so far as my experience goes, it is more like skinning than shaving; the razors are horrible things, as one of the attendants said to me "he should get it off unless the handle of the razor broke." I then understood the saying, "that it is easier to skin than to be skinned."

While confined in my prison-house my mind was continually haunted with the "Lament of Tasso," and that the outside world may have a faint idea of my feelings while there, I will append a few extracts from that work:[D]

[D] Lord Byron, in his travels, found in the library at Ferrara the letters of Tasso, and saw the cell in the hospital at St. Ann's, where Tasso was confined. His enemies charged him with insanity, and threw him into this prison. The manner of treating insane persons in the Old World has been awfully cruel, so far as history gives any clue to the subject. Byron's Lament of Tasso is, no doubt, correct; but this is no reason why in this enlightened age, in a Christian country like ours, that lunatics should be treated as you would treat a mad dog or mad bear.

"Long years of outrage, calumny and wrong; Imputed madness, prison'd solitude, And the mind's canker in its savage mood, When the impatient thirst of light and air Parches the heart; and the abhorred grate, Marring the sunbeams with its hideous shade, Works through the throbbing eyeball to the brain, With a hot sense of heaviness and pain; And bare, at once, captivity displayed, Stands scoffing through the never-opened gate, Which nothing through its bars admits save day And tasteless food, which I have eat alone, Till its unsocial bitterness is gone; And I can banquet like a beast of prey, Sullen and lonely, couching in the cave, Which is my lair, and it may be--my grave. All this hath somewhat worn me, and may wear, But must be borne. I stoop not to despair; For I have battled with mine agony, And made me wings wherewith to overfly The narrow circus of my dungeon wall.

I weep and inly bleed, With this last bruise upon a broken reed. What is left me now? For I have anguish yet to bear--and how? I know not that, but in the innate force Of my own spirits shall be found resource. I have not sunk, for I had no remorse, Nor cause for such--they called me mad--and why? Oh, my judges! will not you reply?

Above me, hark! the long and maniac cry, Of minds and bodies in captivity, And hark! the lash and the increasing howl, And the half inarticulate blasphemy! There be some here with worse than frenzy foul, Some who do still goad on the o'er labored mind, And dim the little light that's left behind, With needless torture, as their tyrant will Is wound up to the lust of doing ill; With these, and with their victims, am I classed, 'Mid sounds and sights like these, long years have passed. 'Mid sights and sounds like these my life may close; So let it be--for then I shall repose.

Feel I not wroth with those who bade me dwell In this vast lazar-house of many woes? Where laughter is not mirth, nor thoughts the mind, Nor words a language, nor even men mankind; Where cries reply to curses, shrieks to blows, And each is tortured in his separate hell-- For we are crowded in our solitude-- Many, but each, divided by the wall, Which echoes Madness in her babbling moods; While all can hear, none heeds his neighbors call-- None! save that one, the veriest wretch of all, Who was not made to be the mate of these, Nor bound between distraction and disease. Feel I not wroth with those who placed me here? Who have debased me in the minds of men, Debarring me the usage of my own, Blighting my life in best of its career, Branding my thoughts, as things to shun and fear? Would I not pay them back those pangs again, And teach them inward sorrow's stifled groan? The struggle to be calm, and cold distress, Which undermines our stoical success? No! still too proud to be vindictive, I Have pardoned tyrant's insults, and would die Rather than be vindictive--yes, I weed all bitterness From out my breast; it hath no business there.

I once was quick in feeling--that is o'er-- My scars are callous, or I should have dash'd My brains against these bars, as the sun flash'd In mockery through them--if I bear and bore The much I have recounted, and the more Which hath no words, 'tis that I would not die And sanction with self-slaughter the dull lie Which snared me here, and with the brand of shame Stamp madness deep into my memory, And woo compassion to a blighted name, Sealing the sentence which my foes proclaim. No, it shall be immortal!--and I make A future temple of my present cell."

TESTIMONIALS.

This is to certify that the Rev. Hiram Chase, a supernumerary member of the Troy Annual Conference of the M.E. Church, resided at Saratoga Springs for one year preceding the spring of 1867; that at the session of his Conference, held that spring, he took an effective relation, and, at the request of the Catharine Street church, Saratoga Springs, was appointed its pastor, and that he faithfully and efficiently discharged the duties of his pastorate--facts, these, which speak for themselves regarding both his mental and his moral status.

SAMUEL MEREDITH,

P.E., Albany District, Troy Conference. ALBANY, N.Y., Aug. 12, 1868.

* * * * *

ALBANY, Aug. 4, 1868.

I have this day listened attentively, and not without as deep emotion as my nature is susceptible of, to Rev. H. Chase's two years and four months in the asylum. I regard said narrative as the unvarnished statement of facts as they occurred during his residence there. I have enjoyed a pleasant acquaintance with the Rev. H. Chase for the last thirty years, and have ever known him to be the same truthful, ingenuous and trustworthy friend, faithful and successful minister of Christ, and a Christian gentleman of more than ordinary culture and refinement. It is an occasion of most devout thanksgiving to Almighty God that he has been mercifully preserved during the past and restored again to his family and many friends, to the fellowship of the church in which he has spent half a century of sacrifice and toil, to her pulpits and altars, and a large place in the best affections of thousands of brethren and fellow-laborers in the church of the living God.

In my opinion the narrative should be printed and widely circulated.

CHAS. DEVOL, M.D.

* * * * *

www.ingramcontent.com/pod-product-compliance
Lightning Source LLC
Chambersburg PA
CBHW060814100426

42813CB00004B/1071